Hazardous
Waste Disposal

NATO • Challenges of Modern Society

A series of edited volumes comprising multifaceted studies of contemporary problems facing our society, assembled in cooperation with NATO Committee on the Challenges of Modern Society.

Hazardous Waste Disposal

Edited by
John P. Lehman

US Environmental Protection Agency
Washington, D.C.

Published in cooperation with
NATO Committee on the Challenges of Modern Society

PLENUM PRESS • NEW YORK AND LONDON

Library of Congress Cataloging in Publication Data

NATO/CCMS Symposium on Hazardous Waste Disposal (1981: Washington, D.C.)
  Hazardous waste disposal.

  (NATO challenges of modern society; v. 4)
  "Proceedings of the NATO/CCMS Symposium on Hazardous Waste Disposal, held in Washington, D.C., in October 1981"—T.p. verso.
  "Published in cooperation with NATO Committee on the Challenges of Modern Society."
  Includes bibliographical references and index.
  1. Hazardous wastes—Congresses. 2. Hazardous waste facilities—Congresses. I. Lehman, John P. II. North Atlantic Treaty Organization. Committee on the Challenges of Modern Society. III. Title. IV. Series.
  TD811.5.N386 1981   363.7'28   82-18050
  ISBN-13: 978-1-4613-3604-4   e-ISBN-13: 978-1-4613-3602-0
  DOI: 10.1007/978-1-4613-3602-0

Proceedings of the NATO CCMS Symposium on Hazardous Waste Disposal, held in Washington, D.C., October 5–9, 1981

© 1983 Plenum Press, New York
Softcover reprint of the hardcover 1st edition 1983
A Division of Plenum Publishing Corporation
233 Spring Street, New York, N.Y. 10013

PREFACE

Putting together these Proceedings has afforded me the opportunity to re-read the papers which were presented last October here in Washington.

In retrospect, the 1981 NATO/CCMS Symposium on Hazardous Waste Disposal maintains the impression of excellence expressed by the participants at the actual event. The scope of the subject-matter and quality of its presentation indicate that we did attain the Symposiums's objective--a comprehensive review of the international status of hazardous waste disposal.

It is my hope that in your own evaluation of these proceedings, you will share my conviction that the Symposium--conceived three years ago, approved by the NATO/CCMS Pilot Study experts in Oslo in October 1980, and taking place a year later--was indeed worthwhile and that this record of its proceedings will be useful for many years to come.

Washington, D.C.   John P. Lehman
June 1982   Editor

CONTENTS

CONTENTS

OPENING REMARKS

Dr. John W. Hernandez, Jr.

U.S. Environmental Protection Agency
Washington, D.C., United States

Good morning. Let me welcome all of you to Washington and to this opening session of the International Hazardous Waste Disposal Symposium. I would like to extend a special welcome to our guest speakers, many of whom have devoted substantial effort for the past eight years to the NATO/CCMS Pilot Study on Hazardous Waste Disposal and its findings to be presented to you this week.

As your hosts, we in the U.S. Environmental Protection Agency are delighted to follow in the footsteps of the Federal Republic of Germany. Germany's excellent leadership during the course of this study has resulted in a series of reports regarded as the most complete compendium on hazardous waste disposal available today.* Those of us who participated in the study have already begun to use the information in our own domestic programs.

The hazardous waste problem is, however, global. It affects not only the industrialized countries such as those of the NATO community. The hazardous waste problem affects countries with developing economies as well. Although their chemical waste output is less in volume, waste management tasks in third-world nations are often complicated by lack of technical ability, lack of resources, and, especially, by lack of information. EPA is pleased, therefore, to see such a broad international contingent here today. It affords us the opportunity to expand and share the NATO/CCMS

---

* See last page for bibliographic citations.

information base. Developing countries, which so often lack the
data necessary for effective planning, can use the NATO/CCMS information as a workable foundation for rapid action. With this data
they can begin to attack their own problems in controlling hazardous
wastes.

In the United States, we regard the issue of improving hazardous waste disposal practices as one of the most serious environmental problems for the immediate future. We are still living with
the tragedy of our Love Canal, as many of you are living with similar examples of poor disposal practices. We are making progress:

o   In February and May 1980, and early 1981, the U.S. Environ-
    mental Protection Agency issued Federal regulations for
    controlling current and future hazardous waste-- from the
    point where the waste material is generated, through storage,
    treatment, and final disposal. This program affects over
    50,000 waste generators and transporters, and about 15,000
    hazardous waste facility operators in the United States.
    Thus, our Nation is now in a transition period as we imple-
    ment this new regulatory program. EPA is monitoring and
    assessing this process, and will streamline procedures for
    these new regulations as necessary to make them work better.
    Also, EPA is working closely with State governments that have
    assumed, or will assume, responsibility for this new program.

o   The United States Congress on December 11, 1980, enacted the
    Comprehensive Environmental Response, Compensation, and Lia-
    bility Act of 1980, the so-called "Superfund." This is our
    country's first attempt to deal with hazardous waste dis-
    posed improperly in the past. "Superfund" sets up a 1.6
    billion dollar fund to clean up problem chemical waste sites
    where those who were responsible for the problem either can-
    not be identified or lack the resources to remedy the situ-
    ation.

As we move forward in the United States, we recognize that hazardous waste disposal is not only a problem which each country is
attacking individually. It is, indeed, a global problem--a transboundary problem. There is not only the point that hazardous wastes
can and do cross national boundaries. There is also the point that
toxic wastes affecting the oceans and the stratosphere--the global
commons--may be controllable only by the world community.

One of the reasons we are meeting together today is the acute
awareness on the part of the Pilot-Study members that continued cooperation is essential if we are to find both--

o Individual solutions, on a nation-by-nation basis, <u>and</u>

o Internationally-consistent approaches to hazardous waste control, when that would be mutually beneficial.

We have with us this week the collective wisdom of most international experts in the hazardous waste field, and I invite you to have an open and frank exchange of ideas. The result will be a productive and rewarding week for all of us. I hope and expect that this spirit of international cooperation concerning hazardous waste disposal will continue throughout all the years to come.

PUBLISHED REPORTS
from the NATO/CCMS PILOT STUDY
on HAZARDOUS WASTE DISPOSAL

	Number	Title	Accession No.
PHASE I	52	Chromium Recycling	PB-279152
	55	Manual on Hazardous Substances in Special Wastes	PB-270591
	62	Recommended Procedures for Hazardous Waste Management	PB-276555
	63	Organization	PB-276559
	64	Landfill	PB-276811
	68	Transportation	PB-279682
	77	Pilot-Study Final Report	PB-286050/AS
PHASE II	118	Thermal Treatment	PB-82114521
	119	Chemical, Physical, and Biological Treatment	PB-82114539
	120	Landfill	PB-82114547
	121	Metal Finishing Wastes	PB-82114554
	122	Final Report	PB-82114562

These reports are available from the US Department of Commerce, National Technical Information Service (NTIS), 5285 Port Royal Road, Springfield, Virginia 22161. Report number 69, "Underground Disposal," is in French and will have to be translated before it can be sent to NTIS.

REMARKS BY CCMS SECRETARIAT

Dr. Robert Chabbal

Assistant Secretary General
    for Science and Environmental Affairs
NATO Headquarters
Brussels, Belgium

EDITOR'S NOTE

Dr. Chabbal was delayed in Europe by NATO business, and was unable to attend the opening session on October 5, 1981. He gave his remarks on October 8, 1981, the last day of the Symposium. In his remarks, Dr. Chabbal reviewed past and present Pilot Studies under the NATO Committee on the Challenges of Modern Society, and provided the context for the Pilot Study on Hazardous Waste Disposal and the Symposium. The text of Dr. Chabbal's remarks is not available.

POLITICAL DIMENSIONS AND IMPLICATIONS OF

HAZARDOUS WASTE DISPOSAL

Dr. Bernd Wolbeck

Federal Ministry of Interior
Bonn, Federal Republic of Germany

INTRODUCTION

Hazardous waste management has developed increasingly to a
central issue in the overall environmental policy in most
industrialized countries. This development reflects the growing
awareness that in no other field of environmental protection one
is faced at the same time with such high quantities and high
concentrations of hazardous pollutants as in the form of solid and
liquid hazardous waste arisings.

Moreover, the dimension of the problem is evident from the
fact that the generation of hazardous wastes is an imminent
consequence of many, if not most industrial activities. Even the
intensive efforts to clean and safeguard the environment in terms
of air and water pollution abatement lead to many and mostly
negative consequences on the waste front. All this demonstrates
that the hazardous waste problem is a permanent and in the long
term probably still growing problem both from the point of view
of generation and disposal.

Hazardous waste management has a high potential for conflicts
of many kinds. This is not only due to the extremely complex nature
of the problem. All measures in the field affect in some way economic activities on the one side and public interests on the other.
This gives rise to strong political implications and often contradictory requests from interested individuals, interest groups, and
lobbies concerned or those groups pretending that they are concerned.
As a consequence, hazardous waste management policy is often a varying and daily compromise between controversial standpoints and interests.

In addition, one has to recognize profound differences in the

institutional and legal systems of individual countries which cause
different approaches to get the problem under control. These factors
to some extent inhibit better international harmonization and cooperation. At the same time, they underline how important it is to overcome such institutional barriers and to get agreement on basic issues,
a result which was achieved by the NATO/CCMS activities.

What is a Hazardous Waste?

The discussion on this question is as long as it is fruitless.
Moreover the discussion confirms my introductory statement that
contradictory requests from the political scene and from the various
parties concerned characterize essentially the answers found to
this question in all our countries.

The issue of how to define and classify hazardous wastes is well
suited to attack or defend bureaucracy stemmimg from environmental
regulations and their enforcement. Just at present we are all aware
in our countries that bureaucracy as a whole is under fire and it
is not a rare observation that the same people and politicians who
formerly proposed complicated and detailed regulations now complain
that these lead to too much bureaucracy. The whole discussion must
be more honest. What I mean, for example, is that the waste generating
industry should not request the regulatory authorities to prepare
highly detailed definitions and criteria for hazardous waste and
at the same time criticize the expenses and administrative burdens
arising from such proposals. Either one is willing to cooperate
with state authorities on a pragmatic basis and then less
'sophisticated' regulatory approaches are possible - as practised
in the United Kingdom, Germany or Switzerland - or one insists on
'perfect' solutions - which do not exist - but then one should not
be surprised about the unwanted side effects as mentioned.

During the past ten years, the political and regulatory discussion has intensively focussed on the question of what constitutes
a hazardous waste. Despite these efforts an international consensus
could not been achieved on this issue. One of the primary reasons
for this seems to me that the question has often been posed without indicating clearly enough the legal requirements which the
definition and classification were to satisfy. A frequent oversight is that the definition of hazardous wastes has to meet quite
different criteria when one looks to the various elements of the
overall disposal system. Requirements for the classification of
hazardous wastes for transportation, disposal or treatment
controls are likely to be more stringent than for purely administrative purposes.

As international discussion on this topic has not been
sufficiently clear, it is not surprising that no common inter-

national definition or classification of hazardous waste exists
and that each of the NATO-countries has its own approach in that
area. Future discussions, therefore, should concentrate more on the
question of the purposes for which some sort of classification is
really needed before continuing the dialog about how a hazardous
waste should be identified.

There are three basic approaches in NATO countries towards the
legal classification of hazardous waste: first, a pragmatic solution
is achieved by describing the waste in a more qualitative way,
indicating type, origin and constituents of the waste; second, a
more scientific approach defining hazardous wastes by certain
characteristics generally involving standard testing procedures;
and, thirdly, definition of the waste in association with concentration limits of harmful substances. In the latter case the presence
of certain listed hazardous components in a waste beyond a defined
concentration makes a waste a hazardous waste.

All three approaches have their value and can be justified.
The first one has the advantage that a legal system, based on that
type of definition, can more easily be administered and enforced
without too much bureaucratic difficulty. It also gives a certain
flexibility to the waste disposal authorities in making qualitative
judgements on individual waste disposal situations and, thus, in
meeting practical needs. The effectiveness of this listing approach
is essentially dependent on the legal system of individual countries,
in particular the flexibility to make case by case decisions by the
controlling authorities to incorporate problematic wastes in the
hazardous waste control system.

The other two approaches have the advantage of presenting a
clear and accurate description of wastes, theoretically leaving
no doubt for the waste generator and waste disposer about how to
deal with the waste. These precise definitions imply a tight
control and surveillance of the definition parameters which in
practice may pose problems with respect to limited manpower, both
on the part of waste generators and the controlling authorities.

Further experience has to be made in future in order to decide
which one or which combination of the three approaches is most
appropriate to ensure proper hazardous waste disposal at minimum
administrative costs. This question must not be seen in isolation
from the overall hazardous waste disposal system. The central
objective of such systems is to dispose of the waste correctly,
and this is less a problem of the definition of the waste than of
actually applying disposal criteria and disposal opportunities.
Until now there is no evidence in practice of a direct correlation
between the quality of disposal and a high degree of accuracy in
the legal definition of hazardous wastes expressed in terms of
chemical composition, concentrations of toxic substances, etc.

Which Controls Are Necessary for Hazardous Waste Disposal?

This issue again is the subject of many controversial discussions
due to the huge number of events which are to be controlled and
the limited personnel with state authorities and private industry
to carry out the control measures. There are figures which tell us
that the expenses for operating a comprehensive control system for
hazardous waste disposal can be a considerable portion of the
overall disposal costs(20 - 40%). It is again the unwanted bureaucratic effort which is opposed to an effective implementation of
control measures. The consequence is that such measures are only
practised and enforced in an unsufficient way, even in cases where
they are required by law. The central task in each hazardous
waste disposal system is, and will continue to be, to ensure that
a hazardous waste is directed along the right disposal route from
waste generator to waste transporters and to a disposal facility,
where the waste may be properly stored, treated, or disposed of.
The surveillance of this waste transfer is perhaps the most difficult
element to manage in the overall waste disposal control system,
as far as the manpower to do it efficiently and the great number
of individual checks which have to be performed are concerned. In
particular, at the time when hazardous waste disposal systems are
about to be established - the present situation in most industrialized countries - these efforts are indispensable.

Without doubt, complete control of hazardous waste flow
raises administrative difficulties. Experience shows that some
bureaucracy is needed to make the whole system work. But on the
other hand it seems that, at least for a certain transitional time,
one will have to accept this effect in order to provide the right
consciousness on the part of waste generators, transporters and
disposers. Like the obligation to submit a tax declaration, the
requirement to use trip-tickets has also an educational effect.
The implementation of trip-ticket systems is a major task in NATO
countries in the future. In doing this, experience will be gathered
which may allow rationalization in the application of the systems.
It is probable after some years of enforcement that the control
authorities will be more familiar with the general movement of
hazardous wastes in their country, so that the control costs
might be reduced.

The transport part of the overall disposal chain is, or can
be, to a greater or lesser extent controlled by incorporating the
transporters in the tripticket system, and/or by requiring a
special license for the transport of waste. The latter requirement
has proved very useful in some countries in selecting reliable
transport enterprises from the great number of interested enterprises
who offer their service.

It is true to say that today the control of existing hazardous
waste disposal facilities, in particular landfill sites, is often

unsatisfactory from an environmental point of view. The operation
of many facilities does not yet meet the quality standards the
existing laws and regulations require. In the interest of the
environment, and because of the long term nature of hazardous
waste problems, concentrated efforts must be launched in all member
countries to tighten the control of hazardous waste disposal, and
to supervise the orderly management of disposal facilities.

In evaluating or introducing control measures in the field of
hazardous waste disposal, it has to be born in mind that such
measures cannot be looked at separately. Necessity, form and
intensity of such measures are directly linked with the general
disposal management concept. Control measures introduced at one
stage of the waste stream control normally influence the control
measures to be taken at consecutive stages. Thus, attempts to
increase the efficiency of hazardous waste disposal control should
never concentrate on isolated aspects but on the whole system.

Improvement of Hazardous Waste Disposal Quality

The intensive discussions within the framework of the CCMS
Pilot Study on Hazardous Waste Disposal have clearly shown that
there exist basic differences in the western industrial countries
in the way hazardous waste disposal is actually undertaken. This
conclusion does not only concern infrastructures for hazardous
waste disposal at different stages of development in terms of
technical equipment and facilities, but also obvious differences
in political attitudes and philosophies about how hazardous waste
disposal should be carried out. The discrepancy between the daily
practice of disposal and official political objectives and
pretensions is obvious.

A priority task in the ensuing years must be to develop a
more common understanding on technical disposal requirements on
the national and international level and to take appropriate
measures that these requirements are actually met. In this context
it is necessary to achieve some sort of harmonization of disposal
standards and quality between neighbouring countries, such as the
European Allies of NATO, or the United States and Canada. This
harmonization is not only desirable for environmental reasons but
also to create a fair and common reference basis for economic
competition for the waste generating industry.

With regard to these objectives, more efforts have to be made
to put hazardous waste management on a sound scientific footing.
Short and long term impacts of various treatment and disposal
methods must be thoroughly investigated and put into perspective
with health and environmental impacts from other human and
industrial activities. The multiplicity of problems concerned
often exceeds the experiences and research capacity of individual

countries. Therefore, close international cooperation and exchange
of information are urgent necessities in this matter.

The international discussion on hazardous waste disposal has
in the past concentrated mainly on political objectives and
regulatory instruments, and measures in terms of directives,
conventions, etc. After this phase of defining the political framework, the time has come to turn from theory to practical solutions
and to work hard on the improvement of disposal itself. Our
generation must avoid creating new "Love Canal" incidents of
tomorrow.

How to Overcome the Siting Problem for Hazardous Waste Disposal Facilities?

The siting of new disposal facilities is probably the most urgent
problem in the area of hazardous waste disposal which NATO countries
face at present. After having introduced strong regulatory control
mechanisms for proper hazardous waste disposal, most of the
industrialized countries experience serious difficulties in offering
waste generators sufficient technical and practical options
(facilities) for them to treat their waste in the way required
by laws and regulation. In many cases there is an obvious discrepancy between regulatory requirements and the actual means of
meeting them. The enforcement gap is not only a permanent threat
to the environment, it also affects the credibility of and the
trust in the quality and usefulness of legal provisions in place.
The central task is to get new hazardous waste disposal facilities
with better technology and operating practices sited and built in
the coming years. There is no other way to solve the existing
disposal problems. Elected officials (politicians) and all other
parties concerned - public authorities, industry, private initiatives, individual citizens - should remember their responsibility
in this respect and overcome the lethargy which to date often
accompanies decision processes in this area.

Elected officials have to tell the truth to the public:
without a positive and constructive attitude to the actual siting
and installation of disposal facilities, the existing laws cannot
really be enforced and the environmental situation will not improve
or will be even worse in the future. This attitude may require
courage to make unpopular decisions, but without courage no major
task in our society can be solved. An ongoing reluctance to take
this attitude for political reasons may increase the uncertainty
amongst the public on the necessity and value of new disposal
facilities and thus stimulate further opposition to them. There
are good examples that quick and politically well timed decisions
in regard to the siting and construction of disposal facilities

are possible and lead effectively to the desired improvement of
the disposal situation. Denmark, France and Germany are to some
extent representative for such developments. In contrast, there
are many other examples which show that without political push and
support private industry and public administration are not able to
solve the problem on their own.

Extensive efforts are needed to inform the public better on
the technical criteria involved in hazardous waste disposal and
on how far precautionary measures are taken to exclude or limit
short term and long term risks. It has to be demonstrated to the
public that disposal facilities are designed and managed properly
so that people feel confident that things are done in the right
way and in the interest of citizens themselves. NATO countries
should cooperate in this field very closely by exchanging information on siting procedures and the operation of disposal plants so
that one country can learn from the solutions found in another one.
There are experiences well proved and persuasive disposal concepts
and technologies existing. What we have to do is to sell them better
to the public.

In this context important questions need further discussion
in the future:

In the overall disposal context should large disposal and
treatment centers be encouraged or should smaller-scale disposal
facilities be the preferred option. The latter solution might in
general be more easily accepted by the local population due to the
fact that the concentration and movement of hazardous wastes is
less in any particular area than in large scale operations. In
contrast, one of the advantages of disposal centers with greater
capacity is that advanced treatment technology can be applied at
reasonable cost and that the overall management and control of the
disposal operations is rationalized.

A second major issue is whether on-site treatment of hazardous
wastes should be preferred to off-site treatment. For many types
of industrial waste production, a greater proportion is treated
on site. On-site operations do take place in the framework of
general industrial production, and so far do not seem to have
extra opposition from local populations. Waste treatment is integrated into overall production processes as a more or less normal
consequence of these activities. But even if on-site treatment
is increased, it can only be part of the overall solution. This
option is usually valid for major waste generators. With medium
and small sized companies the situation is more difficult for
on-site treatment, and for economic and environmental reasons off-
site treatment should be the usual and preferred option.

Hazardous Waste Disposal in the Context of Overall Environmental
Protection

Experience in the past has shown that environmental protection
has been understood mainly to date as oriented towards the
environmental medium to a greater or lesser extent. For example,
air and water pollution abatement policies have often been
considered without sufficiently enquiring how specific measures
for protecting one medium, air or water, would affect the other
one and which new waste problems would be generated.

This sectoral approach has often led and still leads to the
situation where pollution problems are partially shifted from one
environmental medium to another one. The limitation of air emissions
causes new solid waste arisings in the form of sludges and dusts,
often containing concentrated hazardous constituents; water
purification means, in general, more waste sludges. In the end,
many environmental measures result in an increase of waste
quantities which are often very difficult to dispose of.

This simple connection following from the fundamental law of
conservation of mass is not yet adequately considered in many
attempts to improve the quality of the environment as a whole. One
negative consequence of this policy is that the protection of the
soil and groundwater has not received the same political interest
as the other environmental media in the past. This picture is now
changing under the pressure of serious soil pollution problems in
nearly all NATO countries.

A sound policy can no longer afford a sectoral approach to
the environmental problem. What is urgently needed is an integrated ecological approach considering air and water pollution
abatement and waste disposal together. Waste disposal must become
a more integrated part of any effort to clean air and water.
Reciprocal measures and standards in the areas of air and water
pollution abatement have to respect not only needs and requirements, but also technical possibilities and practical limitations
of waste disposal. Otherwise, the obvious lack of enforcement in
the area of hazardous waste disposal can be hardly overcome. This
request is not aimed at reducing the quality of hazardous waste
disposal as one may believe. It is only an argument to get a more
balanced view of the environmental problem as a whole. Environmental policy must not allow that specialists and environmental lobbies
in one area solve their problem at the expense of perhaps more
serious consequences in another one.

In the area of hazardous waste disposal itself an ecolog-
icaly more general view of the problem and possibilities to solve
it is necessary. A striking example is the ongoing political

discussion about the role High Sea Disposal has to play in an
overall hazardous waste disposal concept. International conventions,
in the first place the Oslo/London conventions, require a drastic
reduction of sea disposal of industrial wastes but in many cases
the question concerning which alternative disposal route on land
should or could be followed instead has not been resolved, and
still may not have been discussed. The different ongoing disposal
practices concerning high sea disposal and the obvious uncertainty
about how to proceed further in that area are a clear signal that
the existing conventions have addressed, but have not solved the
problem.

From the environmental point of view, it is not a satisfactory
solution simply to ban sea disposal without offering or indicating
alternative sound disposal or treatment options on land. Efforts
to reduce High Sea disposal have to be based on a well balanced
evaluation of the marine and land environment and the impacts waste
disposal has on the respective media. A sound policy should not
confine itself to saying what should not be done but should above
all indicate what should be done.

In the light of a comprehensive approach to the hazardous waste
problem also the role of recycling and the application of low waste
technologies has to be underlined. This role must gain increasing
importance - not only because of the contents of valuable materials
in many wastes but also as a direct consequence of the pressure
coming from more stringent requirements for hazardous waste disposal
and the limited availability of high quality disposal facilities
to meet these requirements. Therefore, in many situations the
prevention of hazardous waste generation, either by recycling or
reduction at source, is the only viable way to solve that part of
the hazardous waste problem in a satisfactory manner.

Recycling is often understood as a purely economic activity
running separately from disposal in the narrow sense of the word.
This consideration neglects the important contribution recycling
can make to reduce the waste problem with respect to both the
quantities and harmful constituents of the waste. What one needs
in the future is the better integration of recycling options into
the overall waste disposal strategy. In this sense, recycling is
not different from waste management; it must be part of it. This
postulate implies a careful review of existing disposal practices
with the scope to move systematically from "simple" disposal
methods to more resource-saving treatment processes which at the
same time reduce the environmental damage. The potential in that
regard is obvious. This conclusion is proven, for example, by the
considerable reduction of High Sea Disposal, which has been
achieved mainly by recycling efforts in a relatively short time
during the past few years.

Transboundary Disposal - Challenge for Cooperation

In an international symposium of this kind the international movement of hazardous wastes from one country to another one is, of course, a subject of major interest. Some people use the term "hazardous waste tourism" for this activity and this label already indicates that transboundary shipment and disposal do not have the best image to date. To be quite frank, even central governmental authorities and experts in the waste disposal area do often not really know what is happening in the scene. This situation must be improved.

An estimate shows that transboundary disposal of hazardous wastes is today a limited but nevertheless a quite common practice in many NATO and non-NATO countries. In principal there is nothing against it, as long as things are done correctly. On the contrary, the interest of the environment and the economics of waste disposal are often strong arguments to practise this way of disposal. But policy in this area must be guided by certain principles:

- Basically, the disposal of one country's wastes in another country should only be practised if no appropriate disposal option exists at justifiable costs in the country of origin.

- Secondly, the import and export of hazardous wastes has to be subject to strict control, by both the importing and exporting country. Waste stream control must not stop at national frontiers but has to follow the usual route from waste generation to final disposal.

- Thirdly, transboundary disposal must not be considered as a means to escape any country's responsibilities with respect to hazardous waste management. Politically, it is unacceptable that one country should solve its waste problems by exporting a large proportion of its hazardous waste to another country. Import and export of hazardous wastes must be based on the principle of mutual help between countries and not motivated by a one-sided shifting of responsibility.

In the future, NATO countries and other countries should develop a common understanding and a set of international guidelines according to which transboundary disposal should take place. These considerations have not only to concentrate on the elaboration of permit and control requirements but also on the evaluation of political, economic and environmental criteria under which transboundary disposal may be useful and acceptable.

A further important task in the future will be for countries to investigate and quantify waste streams passing their frontiers,

and to make this information available to their neighbours and to
other interested countries.

How to Proceed Further?

In my previous remarks I could only highlight some of the most
important political issues in the context of hazardous waste
management. But already this short insight into the present
situation and problems made quite clear that many questions still
wait for a satisfactory answer, that experience is lacking in
many cases, and that there is still a lot of work to do in technical,
regulatory and, most of all, political aspects. The complexity of
the issue, the transboundary character of the problem and the
very many individual evaluations which have to be carried out should
give rise to a rational sharing of tasks among our countries. We
should learn from each other, the Europeans from the Americans and
viceversa, we should help each other and, in this sense, we should
keep good contacts with each other.

OVERVIEW OF NATO/CCMS PILOT STUDY: DISPOSAL OF HAZARDOUS WASTES

Dr. Bernd Wolbeck

Federal Ministry of Interior
Bonn, Federal Republic of Germany

INTRODUCTION

As with other environmental issues, the hazardous waste
problem started to meet increasing political interest at the
beginning of the last decade. In addition to the general environmental movement, hazardous wastes raised the public attention by
regular press news about new scandals of unproper and illegal
disposal practices and their threat to human health and the natural
environment. More or less all industrialized countries had the
common feeling at that time that something had to be done against
a further mismanagement of hazardous wastes. Governments of nearly
all NATO countries recognized this challenge and responded by
preparing or reviewing legislation and regulations to control the
flow and treatment of hazardous wastes. It was against this background that the Federal Republic of Germany proposed to launch
the Pilot Study 'Disposal of Hazardous Wastes' at the Fall 1973
plenary session. This proposal was received with particular
interest among NATO Allies and other international organizations.

Time Table and Objectives

The Pilot Study was approved by the NATO Council in November
1973. The Federal Republic of Germany acted as the Pilot Country,
the United States as Copilot.

The actual work programme was performed in two phases.
Phase I started in November 1973 and ended in October 1977 with
the submission of a formal final report to CCMS. Then, at the
Fall 1977 plenary session CCMS agreed that the Pilot Study on the

Disposal of Hazardous Wastes should continue into a second phase for
another three years. Based on the results of an Experts meeting con-
vened at Bonn in September 1973, the general objectives of the Pilot
Study were defined as follows:

- to define and analyse technical, organizational and regulatory
  ways and means for the proper disposal of hazardous wastes;

- to compile and evaluate various management approaches to the
  solution of the hazardous waste problem;

- to draw conclusions from the previous analysis and work out
  recommendations where possible.

     The Pilot Study was not limited to technological and
scientific problems as are many other CCMS projects. Basic policy
issues of waste management also were discussed as were economic,
legislative and regulatory aspects and measures as framework
conditions for concrete actions in practice. In this way, the
Pilot Study was able to analyse and clarify a great number of
complex interrelations between technical and non-technical factors
influencing the hazardous waste problem.

     After eight years of intensive work, the conclusion that the
Pilot Study is thus far the most comprehensive and effective
international effort in the area of hazardous waste management
seems to be justified. In the framework of the Pilot Study programme,
nearly all important topics of the problem were investigated. The
results have been summarized in 12 official CCMS documents altogether which give valuable orientation and guidance to national
and international hazardous waste management authorities and
policymakers.

Number and Contents of Pilot Study Projects

     The Pilot Study group selected specific problem areas for
discussion and agreed on a well-defined working programme.
Altogether ten individual projects were chosen:

Phase I (1973 - 1977)

- landfill research and landfill practice
- underground disposal
- transportation
- organization
- recommended procedures for hazardous waste management
- chromium cycle

Phase II (1978 - 1981)

- thermal treatment
- chemical, physical and biological treatment
- landfill research (continuation of phase I)
- metal finishing wastes

Each of the projects was implemented as a self-contained effort receiving inputs from different participating countries. Although there was a certain link between some topics of different projects, the work-programmes of the individual projects were more or less independent of each other. Nevertheless, due to the close cooperation of the members of the Pilot Study group, conclusions reached in the projects tend basically in the same direction whenever more general questions were addressed.

Special emphasis was put in the work programmes on gathering information on the state of the art in NATO-member countries with regard to the topics defined. By collating and evaluating this information, a broad spectrum of activities and various options in the field of hazardous waste was presented. A particular merit of the Pilot Study is that it improved considerably the mutual knowledge of international developments in the hazardous waste disposal area.

It is important to note that the Pilot Study group faced basic difficulties in making general recommendations in the problem areas dealt with in the Pilot Study. Solutions to the hazardous waste problem depend by their very nature on particular national circumstances, e.g., constitutional structures, legal situation, density of population or geological and climatic conditions. These factors vary considerably among the participating countries. Moreover, many questions raised within the projects often do not fall under the responsibility of central government but have to be decided at a more local level according to the specific circumstances involved. Therefore, the output of the Pilot Study is characterized more by a set of different and flexible alternatives than by uniform and ubiquitous recommendations. Precisely because of this diversity of suggestions, the results of the Pilot Study should serve as a valuable tool for practical decision-making in member countries, and in the international community at large.

Participation

The following nine NATO member countries were actively involved in the work-programme of the Pilot Study: Belgium, Canada, Denmark, France, Netherlands, Norway, United Kingdom, United States and Federal Republic of Germany. In addition, the Commission of the European Communities made written contributions to the Pilot Study.

Italy, Portugal and Turkey sent observers to some expert meetings as did the Organization for Economic Cooperation and Development (OECD), Japan and Mexico.

In all, twelve international expert meetings have been held by the Pilot Study group since 1973, three in the US/Canada and nine in Europe.

Approval of the Results of the Pilot Study

The final report of the Pilot Study was submitted by the Federal Republic of Germany to the CCMS Plenary meeting in Brussels May 5, 1981. This Pilot Study report contains altogether 84 major conclusions and 50 recommendations in respect to hazardous waste disposal. At the occasion of its May plenary meeting the NATO Committee of the Challenges of Modern Society (CCMS) approved the following resolution:

NATO member countries do resolve:

(1) to ensure a wide dissemination on the findings of the project reports of the Pilot Study to national governments, local authorities, private industry, scientific community and international organzisations

(2) to pursue the fruitful exchange of information regarding hazardous waste disposal

(3) to consider the conclusions of the Pilot Study projects as guidance to solve the hazardous waste disposal problem

(4) to use their best efforts to implement the specific recommendations resulting from the projects of the Pilot Study

(5) to continue and promote close cooperation amongst themselves by establishing a standing committee of governmental experts

July 3, 1981 the chairman of the CCMS submitted the Pilot Study Summary Report to the NATO Council. The Council took note of the report, endorsed the aforementioned resolution and invited the attention of member countries to the following basic principles for hazardous waste disposal. On these principles agreement was achieved by the participating countries in respect to an effective and integrated hazardous waste management programme:

Basic Principles for Hazardous Waste Management

- Management of hazardous wastes means the organized, systematic channeling of wastes through pathways to assure their recovery or appropriate disposal with acceptable public health and environmental safeguards.

- Fundamental to any hazardous waste management planning are attempts to reduce wastes at the source through process changes, material substitutions and modifications in light of appropriate economics. To the extent that wastes are already minimized, separation and concentration of hazardous wastes are often important in order to reduce special handling and management requirements and increase reuse opportunities. Such concentration will occur as air and water pollution control systems remove noxious pollutants from waste streams.

- An important option in hazardous waste management is the concept of a hazardous waste clearinghouse, that is, to use the waste as is (if possible). One man's waste can be another man's feedstock. Minor process modifications enhance the possibilities for exchange as well. The clearinghouse concept is already being utilized in more than 13 countries including Austria, Belgium, Canada, France, Netherlands, Norway, Sweden, Switzerland, Denmark, Finland, Federal Republic of Germany, United Kingdom and the United States.

- As some hazardous wastes contain valuable basic materials, material recovery makes sense as a next step from both resource conser- vation and environmental viewpoints. As material shortages and regulatory controls become more widespread, material recovery from hazardous wastes will become more economically attractive. If the hazardous wastes cannot be used or materials recovered from them and if they can be safely burned, destruction by incineration with energy recovery would be a desired alternative with due consideration for air emission controls and residual disposal. Traditional land-based incineration as well as ship- board combustion of wastes are options considered viable in several countries. Nonburnable wastes should be detoxified and neutralized by physical, chemical, or biological treatment, again when necessary and feasible, to reduce special handling and management requirements and minimize the amounts of hazardous materials destined for land disposal. The need for such treatment may be a function of economic constraints and local conditions, such as the availability of suitable landfill sites, in certain countries.

- Controlled landfill forms an essential element in the overall disposal strategy. For hazardous wastes disposed of by landfill, volume reduction to minimize land use requirements and to facilitate disposal may be desirable. Isolation techniques (such as waste encapsulation) may be necessary prior to land burial in specially designated landfills in order to assure environmental

integrity of the landfill site. Some landfills may require
natural or artificial liners under disposed waste to prevent
contamination of underground waters. Even if incineration or
chemical, physical and biological treatment options are used,
landfills will be needed for the residues of these processes.

- Some countries find that the use of mines (isolated underground
  areas) or deep injection wells can be reasonabel disposal options
  for selected wastes.

     Each country has a somewhat different view as to the emphasis
to be placed on any of the succession of disposal options. Also
the options should be considered in the light of the quantity
and concentration as well as the type of the waste.

Follow-up Activities

     As was already pointed out, the present hazardous waste
disposal scene is characterized by a rapid development. The
participating countries of the Pilot Study agreed to keep track
of this development in the framework of follow-up activities.
For this purpose an effective reporting and information exchange
system shall be organized, looking after ongoing developments in
the area of hazardous wastes including activities in other
international organizations. A standing committee is to be created
composed by members of the Pilot Study group and other experts
nominated by Governments of NATO countries.

     The terms of reference for this committee should be as
follows:

- to review at regular intervals how conclusions and recommendations
  of the Pilot Study have been implemented;

- to evaluate the progress achieved in the enforcement of hazardous
  waste programs in NATO countries;

- to serve as a forum for the exchange of technical and scientific
  information and national initiatives and programs in the
  hazardous waste management field;

- to facilitate the establishement of compatible waste identification
  methodology through the standardization of definitions and
  criteria for identification of hazardous wastes;

- to identify technical and scientific information gaps and assist
  in the development of priorities of research and technology
  development needs;

- to identify policy and major program concerns which require international discussion.

The standing committee should convene every twelve to eighteen months after the completion of the Pilot Study, in one of the NATO countries.

Dissemination of Pilot Study Reports

In order that NATO countries derive the maximum benefit from the work done within the Pilot Study it is intended to ensure a wide distribution of the 12 project reports. The NATO CCMS Secretariat is negotiating with a private enterprise on a world-wide commercial distribution.

HAZARDOUS WASTE LEGISLATION /ORGANIZATION

Bert A. Szelinski

Federal Environmental Agency
Berlin, Federal Republic of Germany

INTRODUCTION

Hazardous waste management is one of the environmental issues which is under fast development.

When the proposal to prepare a NATO-CCMS Pilot Study on Hazardous Waste was accepted by the NATO Council in fall 1973 the problems connected with hazardous waste management were still largely unsolved but subject to current discussion in most of the NATO member states.

Already at the beginning of its activities towards a pilot study on hazardous waste, NATO realized that hazardous waste management is more than a plain question of applying or developing hazardous waste treatment and disposal techniques. It has also to consider basic principles of environmental policy and its proper place within an overall concept of improving the state of the environment.

Furthermore, technical matters cannot be assessed without considering the framework of conditions and basic aims of hazardous waste management policy. If a country, for example, plans to ban the production of PCB, there will not be a need to spend much time on establishing a disposal concept on this type of waste, provided that there are disposal means available for the interim phase in which these wastes will still be generated. A political approach to change production technology towards non-waste or low-waste technology will have similar effects on waste management.

NATO has seen this interdependence of hazardous waste management policy issues and the technical requirements for waste disposal. One of the basic aims of the phase I sub-project on "Organization" was to work out guiding principles for hazardous waste management which should be considered

27

within the scope of the other, more technical projects, too.

The report on "Organization" was finalized, approved and published in 1977. As said initially, hazardous waste management is constantly developing. It did not stagnate after the report was published. Still, it is safe to say that the conclusions and recommendations as well as the evaluating part of the report are still valid. There are, however, new developments to be reported. Problems which could not be anticipated in 1977 can be clearly seen to-day. With regard to some of its findings the report took a rather optimistic view underestimating difficulties that hazardous waste management has to overcome.

It seems therefore adequate to not only report the findings and activities of the past at this occasion but also to try to follow up and evaluate new developments and trends. Whenever the basis of the report is left this will be indicated in the text. Evaluations and conclusion will then only be the personal opinion of the author and not agreed within NATO-CCMS.

To avoid misunderstanding it is important to make a few remarks on the scope and limitations of the sub-project on "Organization".

There have been sub-projects on "Recommended Procedures for Hazardous Waste Management" and on "Transportation" which cover questions that are of relevance for the "Organization" sub-project, too. To avoid duplication of work and inconsistency of the overall report, the work on organizational matters did not cover definition and classification of hazardous waste although it is quite clear that this issue is extremely relevant for all organizational questions. Definition and classification were discussed within the sub-project on "Recommended Procedures". It has to be borne in mind that evaluations of organization and legislation are made under the presumption that an adequate definition of the term "hazardous waste" as a basis for management decisions is available.

The problems connected with "long-term care" requirements for disposal facilities and/or abandoned sites had to be more or less disregarded in the report on "Organization". They are, however, relevant with regard to control, liability and financing issues and with regard to legislation. They will be discussed later in this paper.

The report on "Organization" concentrated on
- basic conditions for organizational measures;
- form of management of facilities;
- planning of hazardous waste disposal systems;
- control of hazardous waste management;
- liability, financing, insurance.

This presentation of the basic findings of the report will give special attention to political and economic implications and restrictions, to international considerations with regard to import and export of hazardous wastes and will try a general assessment of the role of legislation in this field.

BASIC CONDITIONS FOR ORGANIZATIONAL MEASURES

The first step toward the buildup of a comprehensive hazardous waste management system should be made on the political level.

The direction in which hazardous waste management should be allowed or even encouraged to develop should be made as clear as possible on the political level. The priority of this task is self evident considering that all measures are interdependent and that one step, once it is taken, will influence subsequent decisions.

Furthermore, it should be emphasised that those in the waste generating and waste disposal industries, and those at every level of state administration who are responsible should know well in advance the orientation of future policy.

It might, however, be difficult to follow this principle as hazardous waste management decisions often were emergency decisions in the past, geared to short-term expediency rather than long-term objectives. Nevertheless there might be many countries which have not yet established comprehensive hazardous waste management programmes or systems for which it might be useful to emphasise some principles.

Definition and Classification of Hazardous Wastes

The lack of sufficiently precise definitions constitutes a difficult problem for the introduction of legislation as well as for the construction and equipping of facilities. The possibility that slow development of organizational systems for the disposal of hazardous wastes is to some extent due to the uncertainty of the existing definition cannot be ruled out.

The definition has a special influence on quite a number of areas. For planning purposes, a classification of hazardous wastes is required for the following points of detail:

- determination of the possible disposal method for the various types of waste;

- overall planning (number, locations and distribution of the facilities);

- need for collection sites;

- special regulations for certain types of waste.

A sufficiently precise classification of the wastes is of special importance for control purposes. It permits, on the one hand, control measures with regard to special-wastes (e. g., preparation of analyses, legal provisions concerning disposal or treatment, utilization of certain wastes) to be strenghtened and, on the other hand, a clear definition also permits control to be restricted to selected problematic wastes. As long as hazardous wastes are not defined in a precise but practicable form, an effective control system must extend to a large variety of non-problem

wastes as well. The reliability of the control system decreases with a growing number of wastes requiring control.

Safety measures during transportation, also require a precise classification of the wastes. Since the wastes are normally not chemically pure,
it is difficult to apply to them the current regulations concerning the transportation of hazardous substances. It has to be said, however, that the
problem of safe transport of hazardous waste should be looked at in broad
terms, i. e., considering the transport of hazardous chemicals in general
instead of elaborating a special safety system for the transport of wastes
because potential dangers are very much the same.

The definition of hazardous wastes finally also affects the problem
of financing if special financing methods are to be introduced for certain
wastes.

It is very strongly felt that no means of elaborating a single definition or classification which could cover all these aspects can be envisaged.
Existing definitions usually aim to cover one or two of the mentioned problem areas only.

POLITICAL PRINCIPLES

It needs to be emphasised that hazardous waste management should
not be looked at with too narrow an approach. Ideally the most effective
way to dispose of hazardous wastes is to avoid their generation or to avoid
as far as possible that they have to be treated and disposed away from
the production line. Thus, priority should be given to measures to avoid
or minimize waste generation by use of non-waste or low waste technologies (waste avoidance). It is often stressed that this will simply be
a question of economics and that industry will develop non-waste or low-
waste technology at the time it becomes economical. Unfortunately it is
not quite as simple as that. First of all, the development of new technology
needs incentives. This incentive is in general the competitive situation
of similar industries. These industries, however, do not compete in achieving environmental quality or to gain an award for clean technology. They
compete in price and quality of their products, sometimes even only in
keeping to their budgets. Hazardous waste problems will only be regarded
as a cost factor. Thus research and development with regard to environmental objectives will only start when there is either cost pressure through
high disposal prices or public pressure which will influence the company's
good will.

The same principles will apply for resource and/or energy recovery
with regard to hazardous waste management through recycling technology
which follows next within the hyrarchy of available political options. Recycling options should be looked at not only with regard to the waste generating facility itself but should consider external uses, too. Priority should
be given to promotion of interlocking recycling systems within industries.

ECONOMIC CONSIDERATIONS

The conclusions and recommendations of the Pilot Study Group with regard to economic necessities, problems and restriction connected to the built-up of a disposal system are very limited, indeed. This was partly result of the fact that the report aimed at avoiding any "ideological" commitment towards one special (single) organizational pattern. Basic matters of concern were subsidies and grants vs. the polluter-pays-principle and the strong feeling that economically sound enterprises are an essential must for a disposal market which guarantees the fulfilment of environmental objectives.

The working group came to the following conclusions:

- in principle there should be no use of general budgetary funds for financing of facilities (either private or public owned);

- based on the polluter pays principle, different methods of financing seem possible,i.a.,

    a)   financing by the operator;
    b)   financing by waste generating industries, also through joint ventures (industry owned facilities);
    c)   financing by public and/or private bodies using a fund supplied with payments from waste producers or other sources (for example product charges);

- prices charged for waste disposal should be based on the disposal costs including investment amortization and long term care; subsidized prices should be avoided.

- Equal price levels for disposal, however, should be the objective.

There are many more facts that have to be considered when looking at the economic considerations of hazardous waste management.

As mentioned before, the decision to dispose of a waste is basically an economic decision unless a waste arises unavoidably and cannot be recycled.

Costs of disposal thus play a very important role. It can be said that high costs for waste disposal will have a positive impact on the development of low-waste technology and recycling processes. The same is true for high costs of energy and raw materials. Waste tyres, although not a hazardous waste, could be taken as an example for this principle. Before the energy crisis in 1973 - waste tyres represented a serious problem for waste disposal, particulary through littering. Now they are collected free of charge and no landfill sees any of them as a market has developed for this "fuel substitute". On the other hand, this influences other uses of waste tyres such as the recovery of raw materials from this waste.

The disposal of waste oil and lubricants was such an urgent issue some 15 years ago, that in Germany a special law was enacted to assure the proper collection and disposal. Waste oil management was subsidized with payments out of a fund levied together with the mineral oil tax. Today high prices are being paid for waste oil of good quality and there are suggestions that subsidies for waste oil disposal should be terminated. Such dramatic changes can only illustrate how a situation will be influenced by different economic conditions. They can, however, not be expected for hazardous waste disposal in general.

It takes time for rising disposal prices to influence a waste generating industry to change its production processes. Separate collection of wastes that need special care or treatment will be the first step to be taken to reduce disposal costs unless there is a chance to mix waste in order to avoid disposal as hazardous waste. If hazardous waste mangement is properly enforced it can be noted that quantities of hazardous waste decrease as result of on site treatment such as dewatering, neutralization, detoxication etc. These facts have to be borne in mind when making surveys on hazardous waste arisings as a basis for hazardous waste management planning.

It would, however, be naive to presume that the cost relationships between disposal costs and costs /benefits of non-waste technology or recycling would be the only factor that decides which alternative is used. It might be true under the presumption that other ways than disposal are readily available. This is, however, an exeptional case. Thus theoretical economic assessment has only limited impact in practise.

There is in fact a great deal of psychology in decision-making within industry, particularly where the technical means to reduce waste generation are not fully developed. The influence of budgettary systems cannot be overestimated. It is easier for the management of a company to use existing budgetary funds for current waste disposal than to make a new fund available for research and development or for investments for new technology. The knowledge about economic performance of recycling processes is often poor. Cooperation between different industrial enterprises aiming at establishing joint ventures to overcome hazardous waste problems economically is generally not very well developed, if at all.

To conclude, the decision to use low waste technology or recycling processes is not made simply in evaluating economic facts. To promote developments in this field needs more than a laisser faire, laisser aller attitude. It needs straight forward principles of environmental policy and their enforcement with the necessary legislation. With regard to the conclusions and recommendations of the report two problems deserve special mention:
- disposal prices;
- long-term-care.

Disposal Prices

Waste streams will be directed to facilities which offer disposal at low prices. Although transportation distances will be a limiting factor for long range transportation, there will be competition between disposal enterprises as long as free choice of enterprise is guaranteed. Thus, the question should be asked whether hazardous waste disposal can be regarded as an activity for which the principles of free market societies can be applied without additional safeguards.

The pilot study group was very reluctant to draft straight-forward conclusions with regard to this problem, which may cause irritation in free market societies. Still, the following facts have to be considered: according to different natural or even legal conditions, the economics of different disposal facilities will never be the same, sometimes they will not even be comparable (cf. landfill operations vs. incineration or chemical, physical, biological treatment and landfilling).

Unless there is some subsidizing mechanism for equalising costly disposal operations, technical differences as well as different natural conditions will unavoidably result in different price levels which might not fit into an overall hazardous waste disposal concept. Large price differences can hamper the establishment of sophisticated disposal practises for hazardous wastes even where these techniques are environmentally necessary. Particular problems will arise where investments for disposal facilities have to come from private industry sources.

There are, however, means to overcome difficulties resulting from different price levels such as:
-   establishment of strict disposal standards for certain types
    of wastes linking type of waste and disposal technique,
-   establishment and enforcement of obligatory use of facilities
    linking waste generators to certain disposal facilities by
    assignment of "catchment areas" to facilities,
-   establishment of uniform price levels through cooperation
    of industry or through price regulations.

Each of these options will have its advantages and deficiencies. They might not fit into the economic systems of individual states. It is, however, a fact that different available options to dispose of identical wastes will lead to the use of low-price options. This can eventually result in closure of sites using sophisticated disposal means and to hazardous waste disposal with a low level of environmental security.

Long Term Care

Post closure requirements for hazardous waste disposal facilities will produce costs that should not be borne by the public. There are different options to assure that sufficient funds will be available for this purpose. The coverage of costs for reutilization and long term maintenance of disposal sites, particularly for landfills, should be cared for by including appropriate conditions in the licence. One viable option would be the built-up of reserves; another the creation of a fund or bond to assure the payment for post closure measures. It has to be borne in mind, however, that such financial guarantees for post-closure works will only cover costs that are foreseeable at the time the landfill is still in operation. They will thus not cover accidental events or large scale pollution caused by operational faults. It has to be considered further that there is not enough knowledge available to assess the time frame for which long term care might be necessary. Financial guarantees of the form described can therefore only be regarded as minimum requirements to ensure that waste disposal enterprises cannot go out of business leaving the burden of costs for remedial actions and long term control and maintenance of the site to the public.

The abandoned site issue constitutes difficulties which are different in terms of costs involved and in terms of actions which might be required compared to other management issues. Although there are estimates for the costs involved for investigations of the "sins of the past" and for clean-up actions which might be necessary, there was no general recommendation how these costs should be cared for. Some kind of a "super fund" might be a solution to the problem of financing clean-ups of abandoned sites. It might even provide coverage for accidental environmental risks arising from hazardous waste disposal operations in the future. The United States have so far been the only state that has tried to find a specific solution to finance this kind of environmental damage.

Although this problem area is not covered within the report on "Organization" it might be appropriate to comment on it, as the source of the problem is a lack of legal provision with regard to liability. Moreover, the problem is not limited to hazardous waste disposal, only, but of general relevance for environmental damages. Regulations on liability for environmental damages generally seem to be inadequate.

The following deficiencies can, i. a., be established:
-   difficulties in proving the cause of a damage,
-   the principle of "remoteness of damages,"
-   the principle of fault,
-   insufficient liability capital, both for private individuals and for corporations,
-   lack or unavailability of insurance coverage for environmental risks as compared with operational and accidental risks.

The establishment of a fund accompanied by strict but limited liability seems, at the time being, the only viable option to assure coverage of environmental risks either during operation or after closure of the sites. Such a fund might also cover abandoned site risks.

Limitation of liability might be a means to solve the problems of obtaining insurance coverage for environmental risks but may raise other difficulties.

PRIVATE HAZARDOUS MANAGEMENT VS. PUBLIC INVOLVEMENT

One of the key issues of the sub-project on organization was the discussion and evaluation of different organizational options for the development of a comprehensive hazardous waste disposal system. It is obvious that this issue has to consider the constitutional, political and particularly politico- economic backgrounds of the individual states concerned. It is therefore not surprising that the pilot study group did not end up with a detailed, straight forward recommendation for a single management pattern. It did, however, work out a number of principles which form a basis for organizational decisions which can be summarized as follows:

- the possibilities of establishing certain organizational forms for hazardous waste management have to consider the constitution, the legal system and the political objectives of the country concerned;

- private initiative should be encouraged by appropriate measures and used in so far as possible;

- government authorities ideally should limit their activities to implementation of legislation, control and enforcement;

- private hazardous waste disposal enterprises will require frame conditions that assure
    - a market for such enterprises;
    - liability and a reliable, uninterrupted service from such enterprises;
    - satisfaction of nationwide needs for hazardous waste management by these enterprises;
    - the implementation of stringent waste disposal standards.

Even if waste disposal is organized on a private basis, it may be necessary for government-owned or government-controlled facilities to be available, if the capacity or performance of the private system proves to be inadequate.

Again it has to be emphasized that the establishment of a hazardous waste disposal system on a private enterprise basis would follow mainly economic principles. It has therefore to be assured by appropriate measures that the environmental interests of the public are taken into consideration.

To be competitive, private hazardous waste disposal enterprises would aim at facilities built, maintained and operated at the lowest possible costs within the framework conditions set by legislation. Experience therefore shows that it is very difficult to establish a private enterprise scheme without guarantees, particularly with regard to the economics of operation. As long as there is a choice between different technical methods for disposal there will be a tendency to use low cost technology which will generally mean low quality technology. This might be against public interests.

On the other hand, using private initiative as opposed to public or government involvement has certain advantages. Private enterprises tend to be more flexible and might find it easier than public bodies to cooperate with waste generating industries. The advantages and disadvantages of a

publicly organized waste disposal system may illustrate best the problem
areas which have to be considered before finding a decision.

a)   Advantages

- The creation of an organization for the disposal of hazardous wastes
  with the participation of public corporations can, although not
  automatically, make it possible to ensure directly the adequate
  utilization of the facilities, and direct influence can be exerted on the observation of certain disposal standards.

- The government is obliged to ensure the orderly disposal of hazardous wastes (elimination of dangers to the general public). This
  obligation would be met if the responsibility for disposal was taken
  over by public corporations.

- The creation of a comprehensive system of private disposal enterprises is generally not possible by law. The influence of governmental authorities is often restricted to the licensing of facilities.
  This allows only restricted control to be exercised. Special legislation on organizational forms (e. g., compulsory mergers of waste-
  generating enterprises, the enforcement and financing of non-
  governmental disposal organizations) would raise complex legal
  questions and would be politically difficult to realize. These difficulties are avoided if disposal is a matter essentially for government-supported enterprises.

b)   Problems

- The decision to entrust the disposal of hazardous wastes to public
  corporations would result in a reduction of private investments.
  The readiness to establish private disposal enterprises would be
  bound to decline.

- The separation of waste disposal and waste generation may give
  rise to problems of coordination and cooperation. As a rule, private
  disposal enterprises and disposal facilities of waste generating
  industries are more flexible in adjusting to changing situations
  than government enterprises. In the case of industry-owned facilities, adjustment of waste disposal to the production processes
  and the technical know-how of the operators are additional advantages.

- The operation of disposal facilities by public corporations may
  lead to political priority conflicts. The pursuit of priorities in environmental policy such as waste reduction and waste utilization
  may be affected (problem: sufficient utilization of the existing
  facilities).

- The willingness of waste-generating industries to develop and utilize
  waste reduction technologies may be diminished if all responsibilities were taken over by public corporations.

- The establishment and operation of facilities for the disposal of
  hazardous wastes by public corporations may be a considerable burden on government budgets. This would apply especially if such facilities had to compete with more cheaply run installations of private
  operators (e.g., processing or treatment facilities competing
  with landfill sites).

The disposal of hazardous wastes by public corporations may therefore
raise a number of problems. On the other hand, it has to be noted that all
countries introducing the organization of hazardous waste disposal with
a strong influence of public bodies first try to achieve a supply of facilities run by private enterprises. The proposals for the disposal of hazardous
wastes with the participation of public corporations made in some of the
participating countries are the result of serious deficiencies which appeared
in connection with disposal by private enterprises or which are a consequence of the lack of facilities for the disposal of such wastes.

It has therefore to be emphasized that although priority should be given
to private initiative, a comprehensive system on a private enterprise basis
will not develop automatically but will require serious attempts or behalf
of the responsible authorities by means of contractual, administrative or
other incentives, or even by legislation.

In addition to the principles mentioned in the report the need for including hazardous waste generating industries in the development of a hazardous waste management system has to be underlined. Participation of
the waste producing industries would guarantee a close link between the
waste generating side of hazardous waste management and its disposal side
which can help to avoid misdirected investments and assure the acceptance
of facilities.

Even with industries accepting and fulfilling their role in the development of a hazardous waste management system, it might be necessary
to establish some publicly owned facilities both for emergency cases and
for wastes which need special treatment and for which a "disposal market"
will not develop. Experience shows that the involvement of public corporations or authorities in the establishment of hazardous waste faliclities and
their operation can lead to a quick development of a comprehensive system
but it can have negative effects on the participation of hazardous
waste generating and disposal industries.

PLANNING

A comprehensive system for hazardous waste management needs planning. Without thorough planning the provision of hazardous waste management facilities will remain accidental.

Although it can be presumed that big enterprises will have both the
know-how and the necessary funds to dispose of their hazardous waste properly, provided that this is enforced by law and sufficiently controlled by

government administrations, the same would not be true for smaller enterprises and would not guarantee a nationwide coverage. Therefore overall
requirements cannot be left to individual companies decisions, even in
cases where existing private disposal facilities are made available to the
general public. This is the background for the pilot-study-group's recommendation that governments should play an active role in the planning process for a comprehensive hazardous waste management, either by regulating planning requirements or by actively carrying out planning themselves.

On the other hand, planning for a comprehensive hazardous waste
disposal scheme should not exclusively be made in government offices.
It needs cooperation from the interested groups of the public including
waste generating and waste disposal industries and the responsible government agencies.

The area of coverage is of particular importance for planning purposes. Considering the complexity of the problems and the large group
of substances for which planning has to be envisaged, planning should
be executed for relatively large areas, i. e., at least on a regional basis.
These general recommendations made by the pilot study group may require some explanation. There is definitely much more that could be said
on planning principles but it has to be borne in mind that planning, in
contrast to other management principles, cannot be regarded in a general way but has to be based on national idiosyncracies not only with
regard to legislative principles but also with regard to economical and
geographical facts.

Hazardous wastes are a group of substances which will not fit easily
into a planning scheme. Types and quantities of waste generated tend
to change in short periods of time and are subject to technical changes
as well as to developments in waste management policy.

Looking at states which have tried to establish comprehensive hazardous waste planning mechanisms the following difficulties seem to exist:

- estimation and prediction of waste arisings seem to be extremely in-
  accurate and difficut. There is a strong influence of more and more
  stringent hazardous waste legislation on the amounts and types of waste
  generated.

- Changing prices in raw material markets and rising energy prices bring
  dramatic changes (cf. the examples given on pg. 5 and 6 above).

- Changes in policy and enforcement of hazardous waste management
  principles and of the "state of the art" with regard to hazardous
  waste disposal can have the same impact on waste arisings as a basis
  for planning processes.

- Hazardous waste planning is a very time consuming process, particularly when participation of the general public is involved. The establishment of a plan can already take years; more years will have passed until facilities come into operation.

This calls for a permanent feed-back between practice of waste generation and disposal and planning activities and should lead to the conclusion that planning can only produce a framework for future hazardous waste management, such as suitable sites for facilities and should refrain from binding evaluations with regard to requiring certain kinds of facilities for a given area.

CONTROL OF HAZARDOUS WASTE MANAGEMENT

The establishment of effective control mechanisms is an essential must in organizing hazardous waste management. To exercise control is an indispensable duty of government agencies or other administrative bodies, irrespectively to which organizational pattern hazardous waste management will follow in the individual state.

It has to be emphazized that control has to be established for all stages of hazardous waste management, including:

- hazardous waste generation;
- storage and transportation;
- treatment and disposal.

Licensing of facilities and activities is regarded as basic means to guarantee that hazardous waste is sufficiently controlled. Licensing of facilities should not be restricted to waste treatment and disposal facilities but should also cover waste generating enterprises and waste storage either on-site or of-site. There was general agreement within the pilot study group on this principle although it has to be mentioned that there were different opinions as to where a residue starts to become a waste. There are countries which would not regard on-site treatment or storage as an activity which should fall under hazardous waste management legislation. It was agreed, however, that these activities should then be controlled under production of goods legislation, i. e., through licensing of production facilities.

Waste Generation Control

With regard to possible control measures, control at the source (i. e., control of waste generation) should be given priority. This approach offers the decisive advantage that it can start at the production processes and thus can assure that the principle of priority for non- or low-waste technology and recycling over waste disposal is applied in the production facility. As for subsequent disposal phases, control at source will provide a survey of all kinds and quantities of hazardous wastes and can thus form the basis for waste stream control.

Although the sub-project on organization did not discuss any technical details, it can be said that, wherever possible, it should be made obligatory that hazardous waste leaves the production plant in a condition
that facilitates further treatment and disposal. Control of production facilities should pay sufficient attention to this principle. On- site storage
should only be allowed to collect sufficient quantities for economic transportation. However, there might be cases where orderly treatment and
disposal can, for the time being, not be offered. In such cases storage
at the production enterprise might be a better solution than storage at
an off-site private facility. Experience shows that this kind of "interim
storage" can tend towards improper disposal. In such cases responsibility should therefore remain with the waste generating enterprise which
should be subject to tight control.

Transportation

Transportation is the most critical point in hazardous waste control.
Transportation control should first of all ensure that wastes reach their
final destination. Additionally, it has to be assured that waste transporters
offer a reliable service in terms of safety. This calls for licensing of transportation enterprises. Licenses should only be issued if reliability, proficien-
cy, financial security and technical equipment can be verified. According
to the evidence of the licensing procedure the applicant may recieve a
license that is limited to transport of certain kinds of hazardous wastes
only. Placing high demands on the financial standing of the transport
enterprise and on its technical equipment means, of course, that the same
standards will have to be applied to waste transportation by waste generators. Ideally licensing of hazardous waste transportation enterprises
would lead to a limited number of well-equipped transport enterprises
and reduce the workload for control authorities.

Hazardous Waste Treatment and Disposal Facilities

Waste disposal facilities have to be licensed both for construction
and for operation. Licensing should also apply for storage facilities. The
license should give the necessary details with regard to operation and
environmental protection. It should include provisions that require restoration or stabilization of the site after closure of the facility, long term
care and the financing of these measures.

Waste-Stream-Control

Whereas licensing of facilities and certain activities in connection
with hazardous waste aims at minimizing the operational risks, it does
not guarantee that wastes which are generated will arrive at facilities
which are fit to dispose of them in an orderly fashion. It is therefore ne-

cessary to establish a control procedure which links waste generation
and waste disposal for the control of waste transfer. Waste stream control can be maintained by a waybill or trip-ticket procedure. The competent control authority should be notified at the beginning and at the
end of transportation to get at least a reasonable control of waste
streams. This procedure, however, does not make outside control super-
fluous. It only automatically provides a systematic flow of data on
waste movements. Still, backed by sufficient knowledge on waste arisings in the different branches of industries, it will prevent improper
dumping of hazardous waste as the control authorities will notice if
certain wastes that occur are not reported. As there is notification by
two parties, this system can only be bypassed if these parties collude.

The amount of waste movements might require electronic data processing of notifications and /or trip-tickets, the documents should therefore be prepared for data processing.

A control system based on both licensing and waste-stream-control
will place a higher risk on improper disposal. It requires, however, existing
facilities for all waste arisings to be acceptable. When properly enforced
and credible, such a system can be regarded as a contribution to soothe
public opposition against hazardous waste disposal facilities.

INTERNATIONAL CONSIDERATIONS

The report on "Organization" includes two recommendations with
regard to international aspects of hazardous waste management. The
first calls for a harmonization of technical and safety standards for
waste disposal on the international level to ease international cooperation; the second recommendation pleads for international availability
of sites of a particularly favourable nature.

If the recommendation on international cooperation and harmonization needs any comment to avoid misunderstanding, it should be stated
that harmonization of course should aim at the technical state of the
art or the best practical means available and that it should not be a
harmonization following the lowest standards for environmental protection
practised in the concerned countries. This recommendation does, of course,
not only aim at facilitating international cooperation but also at comparable conditions in economic terms. It has to be borne in mind that
costs of hazardous waste disposal can be an important factor for product
price calculation and it is often, sometimes even reasonably, argued
by industry that certain measures of national governments would influence
the competitiveness of national industries at international markets. Although different socio-economic and geographic conditions may require
different measures for environmental protection, there is some truth
in such an argument. To give a practical example: for titanium dioxide
production it does, of course, make a difference whether waste acids
are discharged into water courses or the sea without further treatment

or are treated or recycled in costly processes. On the other hand it has to be noted that technical development in the field of environmental protection can represent technical know-how which can be favourably used by the industries concerned and that in general, there is only a time lag before such measures are implemented by other countries, too.

The recommendation for transfrontier waste disposal (i. e., import and export) is given under the presumption that international harmonization is sought. It further requires close cooperation of the states concerned and due regard to the cooperative principle, which is a basic principle of international law.

In the last few years in many countries a strong movement against "transborder trade" with wastes has developed. Presenting the report on organization should be the right occasion to comment on that. The pilot study group did not consider transfrontier waste disposal as a moral question. There can, in fact, be certain situations that call for waste export.

The moral question can only be raised when the recipient country has no means to dispose properly of the waste in question.

Waste export will only take place if one or more of the following conditions exist:

- the country of origin has no suitable facilities for the waste in question;

- there is a cost difference in favour of the importing country;

- there are no restrictions for export of waste in the country of origin;

- there are no restrictions for import of waste in the country of destination.

Problems may occur if the country of destination executes no control on the import of wastes. In this case wastes might be imported that cannot be disposed of properly and might cause severe problems.

There is no use for the prohibition of transborder disposal of wastes in general. The duty of the state of origin should be limited to declaration and notification of the wastes in question. This does not need to be a duty of government agencies but should be required from the waste importer. There are strong and convincing reasons in favour of the opinion that it should be a duty of the recipient state to ensure by appropriate measures that wastes will only be imported if proper disposal can be guaranteed. To this end, licensing of the importation of waste and notification from the state of origin, including the necessary information for an assessment of disposal options, can easily be obtained from the country of origin by means of import-licensing. In any case it should not be the duty of the state of origin to assess the disposal options in the recipient state. This might in extreme cases even be regarded as an involvement in the internal affairs of this state. Therefore the report on Organization which is still up to date with regard to this point does not call for any particular method of control of transborder-disposal of hazardous wastes.

LEGAL REQUIREMENTS FOR HAZARDOUS WASTE MANAGEMENT

The legal requirements for a hazardous waste management system can be summarized as follows:

- definition of hazardous waste;
- provision of binding hazardous waste management principles;
- provisions for the distribution of responsibility;
- regulations on financing, including provisions for abandoned sites and long term care and maintenance;
- planning regulations;
- legislation on control mechanisms, regulating
  - waste generation;
  - waste transportation;
  - import of wastes;
  - waste disposal facilities including facilities for storage;
  - waste stream control.

The only prime responsibility for government is the enactment and enforcement of legislation and the implementation and execution of control.

The development of a management system does not need government involvement per se; provided that legal principles can be enforced in practice. Experience shows, however, that even the establishment of hazardous waste treatment or disposal facilities can become a field which needs government involvement either with regard to financing or even to supplying facilities established, maintained and operated by public bodies where private initiative proves insufficient.

The influence of legislation in the field of hazardous waste management should not be underestimated. Still, it has to be borne in mind that legislation as such will not provide the necessary changes immediately. It is, however, indispensible to provide a framework for the enforcement of management principles and at least gives guidance for the concerned parties in which direction developments will have to take place.

HAZARDOUS WASTE DEFINITION AND RECOMMENDED PROCEDURES

John P. Lehman

U.S. Environmental Protection Agency
Washington, D.C., United States

INTRODUCTION

This paper presents the findings of the project Recommended
Procedures for Hazardous Waste Management which was part of Phase I
of the NATO/CCMS Pilot Study on Disposal of Hazardous Waste.[1]
The project was conducted from 1973 to 1977, and was led by the
United States and Canada. The project topics included:

o   Integrated Hazardous Waste Management Program
o   Hazardous Waste Definition
o   Waste Sampling and Analysis
o   Control Measures and Licensing
o   Site Selection and Citizen Acceptance
o   Facility Safety
o   Long-Term Care

The Pilot Study findings for each topic are reported in turn, followed by a discussion of more recent information, where appropriate.

AN INTEGRATED HAZARDOUS WASTE MANAGEMENT PROGRAM

The Pilot Study participants believe strongly that certain
fundamental concepts and options underlie an effective national
hazardous waste management program, and that these concepts and
options are interrelated and must be viewed as an integrated whole
when designing and implementing such a program. These fundamental
principles provide a framework for discussion of the other topics
reported here, and for the other Pilot Study projects as well.

45

Exhibit 1, An integrated Hazardous Waste Management Program, describes an idealized flow of options for managing hazardous wastes from the point of generation through a variety of reduction, treatment, and recovery options to ultimate disposal. Each country has a different view as to the priority and emphasis to be placed on any of the succession of options. This view is based upon the set of cultural, economic, socio-political, and environmental traditions and attitudes held in that country. The Pilot Study countries, however, feel that all of the options described should be mentioned in order to portray the scope and range of opportunities that exist for managing the hazardous waste problem.

The ordering of options in Exhibit 1 is also of significance. The conscientious examination of the processes that generate waste in order to minimize the quantity of hazardous residuals, the transfer of waste materials as raw materials, and the careful consideration of material and/or energy recovery potentials reflect not only concerns for environmental and public health, but also a resource conservation ethic. It is after the consideration of these conservation options that the participating countries endorse the exploration of environmentally sound treatment and disposal options which are technologically feasible and economically sound for the country considering them.

While not addressed in Exhibit 1, it should be mentioned that storage of hazardous waste in containers, tanks, waste piles, or surface impoundments (ponds, lagoons) prior to recycling, incineration, treatment, or disposal, and control of transportation of hazardous waste (such as the use of waste transport manifests or trip-tickets) are significant factors in an integrated hazardous waste management program. Also, some countries have acknowledged that injection of liquid hazardous waste under pressure into deep wells, and land spreading of hazardous waste followed by microbial destruction of organic components, can be effective hazardous waste management options if carried out under appropriate conditions.

HAZARDOUS WASTE DEFINITION

Central to any program to manage hazardous waste is development of a definition scheme which separates those wastes controlled by the management system from the universe of nonhazardous wastes. These definitions are designed to suit the nature of the environmental problem to be solved as well as the socio-political and economic conditions existing in each country.

Exhibit 1
AN INTEGRATED HAZARDOUS WASTE MANAGEMENT PROGRAM

Management of hazardous wastes means the organized, systematic channeling of wastes through pathways to assure their appropriate disposal with acceptable public health and environmental safeguards.

Fundamental to any hazardous waste management planning are attempts to reduce wastes at the source through process changes and modifications in light of appropriate economics. To the extent that wastes are already minimized, separation and concentration of hazardous wastes are also important in order to reduce special handling and management requirements and increase reuse opportunities. Such concentration will occur as air and water pollution control systems remove noxious pollutants from fluid waste streams.

An important option in hazardous waste management is the concept of a hazardous waste clearinghouse, that is, to use the waste as is (if possible). One man's waste can be another man's feedstock. Minor process modifications enhance the possibilities for exchange as well. The clearinghouse concept is already being utilized in more than 13 countries including Austria, Belgium, Canada, Denmark, Finland, France, Netherlands, Norway, Sweden, Switzerland, Federal Republic of Germany, United Kingdom, and the United States.

As some hazardous wastes contain valuable basic materials, material recovery makes sense as a next step from both resource conservation and environmental viewpoints. As material shortages become more widespread, material recovery from hazardous waste will become more economically attractive.

If the hazardous wastes cannot be used or materials recovered from them, and if they can be safely burned, destruction by incineration with energy recovery would be the next desired alternative with due consideration for residual disposal. Traditional land-based incineration as well as shipboard combustion of wastes are options considered viable in several countries. Nonburnable wastes should be detoxified and neutralized by physical, chemical, or biological treatment, when necessary and feasible, to reduce special handling and management requirements and minimize the amounts of toxic materials destined for land disposal. The need for such treatment may be a function of economic constraints and local conditions in certain countries.

For hazardous wastes requiring land disposal, volume reduction to minimize land use requirements may be desirable. The use of mines (isolated underground areas) can be a reasonable method for disposal of selected wastes. Isolation techniques (such as encapsulation) may be necessary prior to land burial in specially designated landfills in order to assure environmental integrity of the landfill site.

In analyzing the issue of definition, the participating countries observed that there were multiple levels of sophistication in identifying what is a hazardous waste. At the statutory level, legislatures and parliaments choose a few phrases to broadly define hazardous wastes and their detrimental public health and environmental effects. Next, at the regulatory level, government agencies further elaborate upon and interpret these broad definitions to produce more specific examples, criteria, categories, or listings of hazardous wastes. Finally, in order to implement these administrative orders and regulations, subnational and local governments and industry must have operational rules, tests, and methods by which to translate regulatory definitions into practice for surveillance and enforcement purposes.

The principal purpose of this section is to briefly examine statutory definitions as background information and then to compare and contrast regulatory definition schemes. Implementation definitions go beyond the scope and level of detail possible in this project. Recommendations as to an "optimal" scheme of definitions are not possible given differences between countries, but conclusions as to similarities and differences of schemes will be attempted.

Statutory Definitions

A number of participating countries passed legislation to control the hazardous waste disposal problem in the 1970 decade. Exhibit 2, Statutory Definitions, highlights the wording of the basic waste management laws in several of the participating countries.

In general, one major theme pervades these statutory definitions. Harmful effects or pollution as the result of a disposal act are the major criteria. Both human health and the environment are specifically cited in France, the United Kingdom, and the United States as possible receptors of negative effects. In Germany and the United Kingdom, wastes with potential hazard are considered "special wastes," and they are part of a larger statutory scheme requiring planning for and notifications of actions relative to a wider range of wastes. In France, the FRG, Netherlands, the United Kingdom, and the United States, chemical, hazardous, or special wastes must be or have been identified, and these wastes are subject to a broad system of regulations.

Thus, all countries begin from a relatively similar statutory basis as to what characteristics of wastes should be of concern. Words such as "noxious," "difficult," "dangerous," "detrimental," "injurious," "poisonous," "polluting," and "degrade" appear throughout the enabling legislation. It is clear that an international consensus exists that certain waste materials--

Exhibit 2
Statutory Definitions

France - . . . categories of waste may be defined by decree and
the enterprises that produce, import, transport or dispose of
wastes which belong to these categories and which are in a state
such that they cause, or at the time of their disposal may cause,
a nuisance such as . . . injurious effects on the soil, plants,
or animals, to degrade the scenery or the countryside, to pollute
the air or water, to create a noise or odor, or, . . . [are] harmful to human health or the environment . . . (Art. 8 and 2; Law
No. 75-633; July 16, 1975).

Federal Republic of Germany - Special wastes are such wastes from
commercial or trade companies which due to their nature, composition, or quantities are especially hazardous to human health, air,
or water, or which are explosive, flammable, or may cause diseases.
Their disposal must be subject to additional requirements according to the Act. (Federal Act on the Disposal of Waste, 1972, as
amended, 1976).

Netherlands - Chemical wastes are: (1) wastes consisting wholly
or partly of chemicals indicated by General Administrative Order
and (2) wastes produced by chemical processes designated by
General Administrative Order. (Chemical Waste Act, 1977).

United Kingdom - Waste "of a kind which is poisonous, noxious, or
polluting and whose presence on the land is liable to give rise
to an environmental hazard." (Deposit of Poisonous Waste Act,
1972); Special wastes are those which "may be . . . dangerous or
difficult to dispose of" (Control of Pollution Act, 1974).

United States - Hazardous waste means a solid waste or combination
of solid wastes, which because of its quantity, concentration, or
physical, chemical, or infectious characteristics may (A) cause,
or significantly contribute to an increase in mortality or an
increase in serious irreversible or incapacitating reversible,
illness; or (B) pose a substantial present or potential hazard to
human health or the environment when improperly treated, stored,
transported, or disposed of, or otherwise managed." Solid waste
includes "any garbage, refuse, sludge from a waste treatment plant,
water supply treatment plant, or air pollution control facility
and other discarded material, including solid, liquid, semisolid,
or contained gaseous material . . . " (Resource Conservation and
Recovery Act of 1976).

whether solid, liquid, sludge, or contained gas--are particularly
capable of damage to the environment and public health and that
controls on their ultimate disposition are required.

Regulatory Definitions

Following establishment of general statutory definitions for
hazardous waste, government agencies typically elaborate upon and
interpret these broad definitions in a regulatory context. This
effort results in more specific examples, criteria, categories,
or listings of hazardous waste to be used by the regulated
community.

In some countries, regulatory definitions for hazardous waste
are developed by the central government. In others, this effort
is performed by State or local government agencies. Sometimes a
combination of these approaches is used, where the central government establishes a basic hazardous waste definition which is then
supplemented by State or local governments.

The term "hazardous waste" is not always used in regulatory
(or statutory) definitions. Alternative terms include "chemical
waste," "controlled waste," or "special waste." Regardless of the
terminology used, the aim is the same: to single out certain
wastes for more careful management control.

The Pilot Study evaluated hazardous waste classification
systems used by Denmark, the EEC, the Federal Republic of Germany,
France, the Netherlands, the United Kingdom, and the United States.
General features of these systems are:

o   They are based on a hazardous waste criteria or
     characteristic approach, a hazardous waste listing
     approach, or a combination of the two.

o   They account for both short-term acute effects and
     longer-term chronic effects of waste components.

o   They do not include radioactive waste, which is typically
     covered under other laws.

While the systems studied share these general features, there are
several variants to the hazardous waste criteria and listing
approaches which merit elaboration.

Hazardous Waste Criteria. In this approach, general hazardous
waste characteristics, such as flammability or corrosivity, are
defined in terms of limiting values of parameters (for example,
flash point or pH) determined by the use of standard test protocols. The waste generator determines via the test protocols

whether or not his waste is a hazardous waste. This approach is
not dependent upon the presence or absence of any particular
substance.

    Listing of Chemicals. In this approach, the presence in a
waste of certain listed toxic chemicals above a defined concentration limit or threshold makes a waste a hazardous waste. This
approach implies that the waste generator must analyze his waste
to determine the concentrations of all of the listed chemicals
using a standard analytical procedure.

    A variant to this approach is used in the United States.[2]
The waste generator is required to test the leachate from a waste
(not the waste itself) for the presence, above specified concentration limits, of certain toxic metals and pesticides. The rationale
here is that the principal concern of a hazardous waste control
system should be those toxic chemicals capable of escaping the
waste into the environment, not just the presence of toxic chemicals in the waste itself.

    Listing of Wastes (as opposed to chemicals). In this approach,
certain wastes are explicitly identified as hazardous waste on a
list prepared by government authorities. No testing is required
by the waste generator since the presence of the waste on the list
automatically brings that waste into the regulatory control system.
The hazardous waste listings must be carefully worded and specific,
however, to avoid drawing wastes into the control system that were
not intended to be included. There are three variants to this
approach:

    o   Listings of "generic" hazardous waste, that is, wastes
         which arise in many different industries or from many
         sources. Examples are: "Waste lubricating oils,"
         "wastewater treatment sludge from electroplating opera-
         tions," or "spent halogenated solvents."

    o   Industry specific waste listings, such as "pickling
         liquor from steel manufacturing."

    o   Specific commercial chemical product lists. These
         commercial chemicals are considered to be hazardous
         waste when and if they are discarded.

Several countries use a mix of some or all of these approaches in
their regulatory definitions of hazardous waste. Each has advantages and disadvantages.

    The hazardous waste criteria approach is perhaps the simplest
in concept. However, the number of hazardous waste criteria or
characteristics for which standard test protocols are inexpensive

and readily available is quite limited. This is especially true
for long-term effects, such as chronic toxicity and carcinogen-
icity. Further, this approach places a testing requirement on
all waste generators, unless it is limited to certain industries
or waste sources.

The listing of chemicals approach has merit in that it is a
relatively straightforward task to identify those chemicals of
principal concern to human health and the environment. It can be
quite difficult, however, to identify and justify a particular
concentration level or threshold for each chemical on the list,
above which a waste is considered to be a hazardous waste. This
is especially true for carcinogens, mutagens, and teratogens for
which no "safe" threshold is widely recognized. As with the
criteria approach, this approach requires definition of standard
test protocols for all chemicals listed, and also requires all
waste generators to test their wastes for all the listed chemicals,
unless somehow limited in scope.

The principal advantage of the waste listing approach is that
waste generators do not have to analyze their wastes. The government agency must, however, define and use criteria for listing
wastes as hazardous wastes. Further, given the limited resources
available in government agencies, it takes a long time to identify
and list all of those wastes which should be considered hazardous,
when a listing approach is used.

Some national, State, or local governments have carried, or
are attempting to carry, the regulatory definition of hazardous
waste one step further by subdividing the universe of hazardous
waste into ranked categories based on the "degree of hazard" of
the waste. Clearly, some wastes are more hazardous than others,
and management controls could be tailored to reflect the degree of
hazard of the wastes at issue. This approach has considerable
appeal in that human health and the environment could be protected
from improper hazardous waste management at the least possible
expense to society. The difficulty with this approach lies in the
fact that the true "degree of hazard" of a particular waste depends
not only on its intrinsic properties (such as the degree of toxicity, corrosivity, etc.) but also on several other factors as well,
including the waste quantity, the type of management scheme selected for the waste, and the location of the waste management facility with respect to population density, climate, and geological and
hydrological conditions underlying the facility, to name a few.

Taken together, the large number and wide range in the limits
of the variable factors involved make it very difficult to define
a "degree of hazard" ranking and companion management standards for
hazardous wastes on a national basis. Nevertheless, the advantages
of a "degree of hazard" system are sufficient to spur development

of the concept in future years, perhaps dealing with general
classes of hazardous waste at first, followed by schemes applicable
to individual wastes.

A complete international consensus on the regulatory definition of hazardous waste is not possible at this time, given the
various definitional approaches selected by each country, and the
differing socio-political and economic conditions existing in each
country. Furthermore, the definition of hazardous waste is a
dynamic activity with countries adding, modifying, or deleting
hazardous wastes from their regulatory control systems over time
as new information becomes available, or as mandated by national
legislatures or parliaments. However, the hazardous waste classi-
fications evaluated during the Pilot Study reveal some consensus
as to which wastes should be regulated as hazardous wastes.
Exhibit 3 illustrates 38 wastes or waste classes which three or
more participating countries consider to be hazardous waste,
according to their lists or definitions of hazardous or potentially
hazardous waste, as they stood in 1977. It is interesting to note
that the majority of hazardous waste classes that three or more
countries have in common are those defined on the chemical-compound
group level.

WASTE SAMPLING AND ANALYSIS

Waste sampling and analysis procedures are the cornerstone of
any viable waste management system. In the "hazardous waste criteria" and "list of chemicals" approaches to hazardous waste definition (see previous discussion), the results of sampling and
analysis procedures determine whether or not a waste is included
in the hazardous waste control system. A later step is to further
analyze the regulated wastes, whether listed or not, sufficiently
to enable proper management in storage, treatment, and disposal
operations.

Wastes can be solids, liquids, sludges, or contained gases.
Some wastes are complex mixtures of sludges and liquids in several
phases. Many wastes contain both organic and inorganic materials.
Therefore, it is clear that several different waste sampling and
analysis protocols are needed to deal with the diverse kinds of
waste encountered.

When this part of the Pilot Study ended in 1977, none of the
participating countries had established standardized, validated
waste sampling and analysis protocols. However, standard analytical protocols developed for materials testing, and air and water
quality programs, were being used successfully for many wastes.
This section summarizes the efforts reported by participating

Exhibit 3
WASTES CONSIDERED HAZARDOUS OR POTENTIALLY HAZARDOUS
BY THREE OR MORE COUNTRIES

Aluminum-Containing Waste

Antimony & Compounds

Arsenic & Compounds

Asbestos

Beryllium Waste

Cadmium Waste

Chlorine

Chromium III Waste

Chromium VI Waste

Copper Waste

Cyanide Compounds

Dye Manufacturing Waste

Fluorine

Halogenated Solvents

Herbicides

Isocyanates

Laboratory Waste

Lead Waste

Magnesium Waste

Mercury Waste

Metal Surface Treatment Waste

Nickel Waste

Non-Halogenated Solvents

Oil Refinery Waste

Organic Peroxides

Paint Manufacturing Waste

Pesticides

Pharmaceutical Manufacturing Waste

Phenol-Containing Waste

Phytopharmaceutical Waste

PCBs

Rubber Manufacturing Waste

Silver-Containing Waste

Sulfur-Containing Waste

Thallium Waste

Vanadium & Compounds

White Phosphorous

Zinc Waste

countries at that time, and comments on more recent developments in the United States.

Waste Sampling

Unless the sample taken is representative of the waste material as an aggregate, the information extracted from the sample is misleading. The need for standardized sampling procedures is obvious.

The Federal Republic of Germany's contribution to the Pilot Study stressed the following points:

a. The heterogeneous composition of waste dictates that all phases must be sampled (i.e., a complete vertical sample of fluid portions as well as solid residue and vapor phase must be taken).

b. Waste production may vary as to the time or length of operation (i.e., effluent off the initial stages of a process may differ from effluent off later stages). The sample taken must be representative of all the effluent.

c. When withdrawing aliquots for analysis from samples, the aliquots must also be representative of the aggregate waste. This may require mixing of the sample before withdrawing the aliquot.

d. Sample preservation methods should mimic the ambient conditions of the wastes' "natural" environment (e.g., anaerobic sludge samples should be protected from air, etc.).

e. Sample fixation may be necessary for wastes containing volatile or unstable components.

The United States noted in 1977 that methods employed for sampling materials with consistencies similar to wastes can be adapted for use in waste sampling. These methods are primarily of two kinds: those that can be used for fluids and those that can be used for granular non-fluid type materials. Examples of the former include the California Department of Health's Coliwasa sampler[3] and the oil thief for non viscous fluids (see ASTM standard, D270-23).[4] Examples of the latter include soil augers (see ASTM standard D1452-19) [5] and grain sampling triers.

In 1980, the United States Environmental Protection Agency established guidelines for obtaining samples of wastes from various environments. These span the universe from the Coliwasa, dippers, and weighted bottles for liquids in tanks, pits, and ponds, respectively, through thiefs, triers and augers for granular

solids. These methods are described in detail in 40 CFR Part 261 of
the U.S. hazardous waste regulations and supporting documents.[3]

Waste Analysis

The analysis of waste materials is complicated by the following factors. The composition of waste constituents often ranges
over a wide variety of chemical types and over many orders of
magnitude of concentration. The presence of other constituents in
a sample under analysis can introduce interference, since wastes
often contain many of these other constituents. It is very difficult to compensate for their presence in the interpretation of the
analytical data. Also, most wastes are not in a physical form or
state that is amenable to analysis. For a standard analytical
methodology to be prescribed, the wastes must be in some sort of
"standard state." For this reason, significant amounts of pretreatment are often required before analysis can begin. The
following is a discussion of how various organizations are addressing the problem of waste analysis. The discussion includes references to analytical procedures applicable to the waste itself,
as well as to waste leaching, extraction, or elutriation procedures prior to analysis of the elutriate for various chemical
species. The "leaching" of hazardous constituents from a waste
by the various liquids with which it comes into contact is one of
the primary vectors of pollution from land disposal. Therefore,
waste "leaching" procedures are very appealing in some countries.

Federal Republic of Germany. The Society of German Chemists,
in its handbook "German Standard Procedures for Analysis of Water,
Effluent, and Sludge," described standard methods for determining
water content, organic content, and pH of waste samples. Methods
for determining free cyanide and mercury content of wastes were
under development in 1977.

The FRG also has a waste leaching or elutriate procedure for
waste prior to analysis. The procedure does not attempt to repro-
duce the exact parameters of natural leachate, but only serves to
determine if various chemical species may dissolve when the waste
is in contact with water. A general outline of the procedure is
given in the Pilot Study report.[1]

Canada. In April 1975, the Solid Waste Management Branch,
Environmental Protection Service, Environment Canada, and the
Department of Civil Engineering, University of British Columbia,
co-sponsored an international round table seminar on landfill
leachate analysis. The Seminar Proceeding Report, "Procedures
for the Analysis of Landfill Leachate,"[6] contains recommendations on analytical techniques which are appropriate for leachate
characterization and analysis. The methods of choice for the
analysis of most chemical contaminants in leachate-type medium

are those contained in "Standard Methods for the Examination of
Water and Waste Water"[7] and the U.S. EPA Manual entitled
"Manual of Methods for Chemical Analysis of Water and Wastes."[8]
The seminar proceedings also, however, mention modifications to
the techniques recommended by these references that they have
found helpful, problems and interferences the seminar participants have run into in using these methods, useful pretreatment
techniques, as well as other techniques not referenced by these
two manuals, which the participants have used with success.

United Kingdom. There are no mandatory standard analytical
procedures in the United Kingdom, nor are there likely to be. It
is the sentiment there that formulating such standards is not what
is needed, due to the difficulties of specifying these standards
and ensuring enforcement; rather, an open dialogue of the various
problems attendant to waste analysis should be encouraged. Furthermore, many of the methods recommended by the United Kingdom's
Department of the Environment for water analysis should also be
applicable to waste analysis, and organizations such as the
Society for Analytical Chemistry have collected recommended
methods for analyzing materials such as trade effluents. Preliminary work is now being undertaken for the Department of the
Environment with a view to commencing a research project on the
leaching of wastes.

United States. In 1977, the United States had no standard
procedures for either analyzing regulated waste materials or for
determining the potential mobility (leachability) of toxicants
present in such wastes. In 1980, however, the U.S. Environmental
Protection Agency established mandatory standard procedures for
both waste analysis and waste leaching as part of its hazardous
waste regulations. These methods include measurements of waste
ignitability based on flash point tests, of corrosivity based
on pH and reaction with steel, of reactivity based on explosivity,
and of toxicant mobility based on the Extraction Procedure. The
Extraction Procedure used to determine toxicant mobility approxi-
mates natural leaching processes by using a mild organic acid
extractant coupled with steps to simulate the movement of the
separate phases of the waste in the landfill environment. In
addition to the aforementioned test methods, a number of analytical procedures were developed and promulgated for use in determining the presence and concentration of a wide variety of hazardous substances that may be present in wastes. These methods are
described in detail in "Test Methods for Evaluating Solid Waste."[3]

The above discussions illustrate that the problem of waste
analysis is far from solved. The large number of potential constituents, their differing properties, the wide range of concentrations they may be present in, as well as the many different
matrix effects, make the problem of developing standard methods

exceedingly complex. There are useable inexpensive techniques
for most inorganics of concern, especially heavy metals; the
interferences involved in these techniques are understood to an
extent, and pretreatment can often overcome these difficulties.
Extensive separation is often required before analysis, and the
interpretation of the analytical data is often ambiguous. Unless
the analyst has some fairly good idea as to what organics are
present in the waste, presently available methods are too expensive
and time consuming to be of use in routine waste analysis.

Elutriate tests presently being used are not precisely comparable to natural leaching action. Natural leachate solubilizes
waste components because of many physical parameters of the leachate including pH, dielectric constant, organic content (which can
act as chelating agents, buffering agents, etc.), temperature,
redox potential, and others. It is impossible to address all these
parameters in all their potential ambient ranges, in any one standard leaching process. Therefore, more work needs to be done in
this area.

Further work has also been done in some countries in connection with other disposal methods, e.g., incineration, in the form
of standard incineration tests and sampling and analysis of slag,
fly ash, and flue gas. However, due to the lack of information on
these methods at present, these considerations were not covered in
this report.

CONTROL MEASURES AND LICENSING

Once identified as a hazardous waste, proper management of the
waste requires administrative control by responsible authorities
from the waste's "cradle to the grave." Thus, the producer or generator of hazardous waste should label the waste containers. Where
hazardous waste is transported, a manifest or trip-ticket system
should be used to ensure that the waste arrives at its designated
management facility. Lastly, all storage, treatment, and disposal
facilities for hazardous waste should be licensed. These control
measures are discussed below.

Labeling

Unlike consumer products or commercial commodities, there is
very little incentive for labeling wastes. Yet, labeling plays an
important role in the proper management of hazardous wastes. Numerous accidents and deaths have occurred because waste containers
had no label or warning signs to alert employees handling them.

All containers of hazardous wastes should be appropriately
labeled. As a minimum, this label should provide the producer's

name and address, and a description of the container contents. If
the waste is to be handled by second or third parties, it is also
desirable that the container of a waste indicate in some manner the
nature and degree of hazard associated with the waste, and any special handling and disposal procedures which may be required.

It should be noted that there is a distinct difference in the
applicability and function of a hazard information label and a
transportation label or trip ticket. The two are complementary and
could possibly be combined. A hazard information label developed
for use in Canada is described in detail in the Pilot Study
report[1].

Transport Manifests or Trip-Tickets

A hazardous waste transport manifest or trip-ticket is a form
of documentation primarily designed to track the waste from generator to transporter to disposer. Most waste manifest systems include
a mechanism for reporting waste transport events to responsible
authorities. Thus, the transport manifest may also generate information for planning and surveillance purposes.

While different laws and regulations have led to various transport manifest systems in different countries, the basic aim is the
same: to monitor hazardous waste flow to disposal facilities. General principles of transport manifest systems are:

- The waste manifest form has several copies.

- Waste generators, transporters, and disposers are registered with responsible authorities.

- The waste generator originates the manifest, provides his
  name and address, describes the waste to be shipped and
  its amount, specifies the destination, and signs the
  manifest.

- The waste transporter then signs the manifest when he
  collects the shipment, and carries the manifest with the
  waste during transport. The generator keeps a signed
  copy of the manifest, and, in many systems, also sends
  another copy to the responsible authorities.

- When the waste arrives at his facility, the waste disposer
  verifies that the waste received is the same as described
  on the manifest, signs the manifest, and gives a copy to
  the transporter. The waste disposer keeps a signed copy
  of the manifest, and, in many systems, sends a copy to
  the responsible authorities.

In most systems, the authorities thus verify that the hazardous waste actually reached its intended destination, and each party has a record of the shipment. The Federal transport manifest system used in the United States, however, places the responsibility for verifying shipments on the waste generator. The waste disposer sends a copy of the manifest back to the generator, not to the authorities. If the generator does not receive this verification of waste arrival, within 45 days from the shipment date, he notifies the Federal authorities. This "management by exception" approach relieves the Federal authorities from dealing with large numbers of routine transactions, and focuses attention on problem cases.

Facility Licensing

There is agreement among countries participating in the Pilot Study that all hazardous waste storage, treatment, and disposal facilities should obtain a license or permit to operate from responsible authorities. Some countries also require a license for facility construction prior to operation. It is not possible to specify one set of license conditions applicable to all hazardous waste facilities due to the many types of facilities, the numerous different classes of wastes handled, the many possible settings for facilities, and the different forms of national and regional legislation and regulations which may be applicable. It is generally recognized, however, that hazardous waste facility licenses or permits should address:

- Facility siting or location factors.

- The types and amounts of wastes which a facility may receive.

- The operating conditions which must be met.

- Equipment maintenance and inspection procedures.

- Proper training for facility staff.

- Control of emissions to air, surface water, and underground water.

- Monitoring of emissions.

- The records which must be kept.

- The reports which must be made to controlling authorities.

- Facility safety procedures.

- Provisions for proper facility closure, and, for land disposal facilities, long-term care after closure, and

- The right of access for inspection by controlling authorities.

Most of these licensing factors are highly dependent upon the type of facility, the type of waste handled, and the environmental setting of the facility. Thus, license conditions must be determined individually, on a case-by-case basis. The Pilot Study did, however, address facility site selection, facility safety, and long-term care of land disposal facilities in more detail. These factors are discussed in the following sections.

The processes used for issuing facility licenses and permits, including the degree of public participation involved, vary greatly from country-to-country, and thus were not addressed in the Pilot Study.

SITE SELECTION AND CITIZEN ACCEPTANCE

Careful selection of sites for hazardous waste management facilities is one of the most important first steps for controlling hazardous waste. Citizen attitudes towards hazardous waste management facilities are almost always negative, but there are methods to deal with this issue. This section reviews approaches to these matters used in participating countries.

Site Selection

The Pilot Study reviewed the approaches used in the State of Baden-Wurttemberg in the Federal Republic of Germany, in the United Kingdom, and in the United States (EPA and California and Illinois), to the problem of identifying hazardous waste site selection criteria. These criteria primarily apply to land disposal facilities; different criteria may apply to other types of hazardous waste facilities, such as container storage, incinerators, and treatment tanks.

A total of 88 different factors or criteria associated with site selection were identified at least once in the five approaches reviewed in this study. These criteria included not only technical factors (such as size of the facility, type of waste, climate, hydrogeology, and proximity to surface waters), but also, in some cases, social and economic factors (such as conformance with land-use plans, proximity to population centers, and transportation access).

Exhibit 4 lists 14 of the 88 criteria which were identified by
three or more approaches. As might be expected, it is the hydrogeological factors which are of most common concern. Consequently,
close collaboration with geological experts is desirable when
choosing sites for land disposal facilities.

There appear to be three basic approaches to site evaluation.
The State of Baden-Wurttemberg (FRG)[9] uses a matrix approach
where all selection criteria, whether quantifiable or not, are
assigned weighted values such that they are all considered as part
of the total decision. The systems used in the United Kingdom[10]
and the States of California[11] and Illinois[12] all depend on the
considered selection of criteria of concern, and the judgmental
interpretation of hydrogeological and other technical, social, and
economic data that are gathered. The U.S. EPA in 1976 proposed[13]
a decision tree whereby a site is first screened for technical
suitability; only technically acceptable sites are then analyzed
for less tangible factors which might influence selection of a
site. Although these systems differ in their site evaluation methods, they share certain site selection factors which include measurable soil and ground-water conditions.

More recently, in January 1981, the U.S. EPA promulgated as
regulations[14] two basic requirements for hazardous waste facility
locations in the United States. These related to seismic activity
and river floodplains. New facilities must not be located within
61 meters (200 feet) of a seismic fault which has had displacement
in Halocene times (roughly the last 11,000 years). Such faults are
located predominantly in the western United States. All facilities
located in a 100-year floodplain must be designed, constructed,
operated, and maintained to prevent washout by a 100-year flood.
This can be done by building dikes, or by other methods.

Citizen Acceptance

Citizen acceptance is a most important factor in hazardous
waste facility site selection--one that is impossible to quantify,
but that canot be overlooked. The community disposition towards a
facility can make the difference in whether or not it is a success.

Contributions from France[15] and the United States[16] to the
Pilot Study described public education programs for assisting
siting of waste management facilities. The degree to which public
education programs are used, and the degree of consideration given
to citizen commentary which usually results, naturally varies
because of the nature of various governments.

Where public education methods are used, government officials
and industry representatives must be open and straightforward in
their presentation of the proposed site to the community. Guidance

Exhibit 4
SITE SELECTION CRITERIA IDENTIFIED BY
THREE OR MORE COUNTRIES

Effect on Drinking and Industrial Water Supply

Effect on Wastewaters and Percolating Water Drainage

Effect on Surface Water Drainage

Imperviousness of Subsoil (permeability)

Water Resources in Surrounding Area

Surface Water Drainage

Flood Protection

Expected Quantity of Percolating Water

Suitability of Soil

Landslide, Earthquake Hazards

Effect on Long-Term Surface Water Drainage

Precipitation

Depth to Water Table

Ion Exchange Capacity of Soil

by professional communication consultants can be helpful in this
process. In no event should it be claimed that a hazardous waste
facility has zero risk to the community. On the other hand, it
can be fairly pointed out that hazardous waste facilities are in
many ways analogous to public utilities or services (such as water
supply, electrical power, and transportation systems) which are
commonly provided to attract industry to a community, thereby providing employment and contributions to the local tax base.

More recently developed methods for gaining citizen acceptance[17] include the use of environmental mediation and State-wide
hazardous waste facility siting boards. Environmental mediation is
adapted from labor mediation methods. An objective third party
intercedes between the proponents and opponents of the site, tries
to resolve issues, and, in some cases, may negotiate compensation
to the community.

State-wide siting boards or advisory committees have been set
up recently in several States in the United States. These boards
typically have members from the local community, industry, and
government. Operating in conjunction with a State-wide hazardous
waste management plan, these boards or committees can ease fears
that a particular site was selected arbitrarily. Some boards have
the power to override local opposition to a site, which may ultimately be necessary despite efforts to involve and educate the community about the proposed facility.

FACILITY SAFETY

In theory, the safety precautions and procedures of a hazardous
waste management facility should be the same as any chemical plant or
laboratory that handles hazardous materials. The fact is, however,
that there is a much greater degree of uncertainty in the handling
of hazardous wastes. The wastes are of no economic value and,
therefore, the most efficient method of handling is to combine a
large number and variety of wastes in treatment, storage, or disposal operations. The success of this approach will depend on the
experience and expertise of the supervisors and handlers of wastes
at these facilities.

In anticipation of unknown hazards such as gas evolution,
toxic vapors, flash fires and explosion, and unintentional reaction,
the personnel at a hazardous waste facility should have on hand
equipment to handle accident situations that would not normally be
anticipated in domestic waste facilities. Personnel should be
trained in first aid and emergency procedures.

Before treating any wastes, the personnel should be totally

aware of the nature of the specific constituents of the waste and
have at their disposal prescribed precautions and an analytical
capacity to verify composition of the waste either on the premises
or elsewhere.

Personnel protection gear should be provided. Depending on
the particular hazards involved, these might include rubber suits,
rubber gloves (light and heavy), hard hats, face shields, self-
contained breathing apparatus, ropes and safety belts, rubber knee
boots with safety soles and toes, first-aid kits, tools for opening
containers (in some cases, remote-controlled) and oxygen resuscita-
tion equipment. When personnel are authorized to work alone or in
pairs, communications should be available at all times.

Local emergency agencies, such as the fire and police departments, should be alerted to the nature of the operations taking
place and be notified of the emergency procedures to be followed in
case of accidents at the facility. The management of the facility
should also initiate a safety program and have an emergency response
team for emergency situations.

It is extremely important that a professional supervisor be
aware of the operations taking place at the facility, and that
personnel do not handle any material or start any process that
they are not entirely familiar with. Therefore, when a facility
receives a waste or contemplates receiving a waste which they are
not familiar with, laboratory and small-scale testing of treatment,
storage, or disposal procedures should be mandatory.

In summary, the technology and techniques for handling hazardous wastes are essentially the same as handling other hazardous
commodities, but the degree of uncertainty is much greater and,
therefore, a proportionate increase in safety precautions should
be taken at hazardous waste management facilities.

LONG-TERM CARE

Many hazardous wastes are essentially indestructible in nature.
Others decay to non-hazardous forms very slowly. When such wastes
are disposed of in the land, e.g., in a landfill, their potential
for harm to human health and the environment does not end when the
land disposal facility is closed. Long-term care of these facilities may be necessary in many cases.

The Federal Republic of Germany, the United Kingdom, and the
United States take the long-term effects of placing hazardous waste
in land disposal facilities into consideration in their hazardous
waste management programs. The FRG and UK programs depend heavily

on their licensing systems to ensure that hazardous waste land disposal facilities are properly located and operated, thus minimizing
long-term effects after closure. The U.S. program goes beyond this
by explicitly placing post-closure technical, administrative, and
financial requirements on the facility operator.

In the FRG system, once a disposal plan is approved, the
problem of long-term care of a hazardous waste landfill site is
assumed to be resolved, since the approved disposal plan includes
long-term provisions for recultivation and reclamation of the site.
In addition, the conditions of a license can be modified by the
supervising authority at any time, or the license can be revoked.

The UK licensing system, like that of the Federal Republic
of Germany, is designed to assure adequate long-term care of a
disposal site. If the site becomes a hazard through a miscalcula-
tion on the part of the waste disposal authorities when granting
a license, then it is anticipated that the matter will be settled
in the court. In principle, liability for damage caused by contra-
vention of license conditions falls to the operator, but the equi-
ties of the situation are ultimately settled by adjudication.

Recent U.S. hazardous waste regulations explicitly address
long-term care of hazardous waste land disposal facilities after
closure.[18] Facility owners must place a notice in the property
deed that hazardous waste is buried at the facility. They must
provide to government authorities a record of the types and
location of hazardous waste buried within the facility. For 30
years after closure, they must monitor ground water beneath the
facility for signs of pollution, and maintain waste containment
and security devices, such as surface water run-on controls,
landfill cover, and fences. Further, government authorities may
place restrictions on future use of the land after closure, in
order to assure that waste containment devices are not disturbed
or breached.

In addition, and perhaps most important, the U.S. regulations
assure that the land disposal facility owner or operator will have
the financial resources for post-closure monitoring and maintenance.
The facility owner or operator must build up a trust fund over the
operating life of the facility sufficient to pay for post-closure
activities, must post a surety bond for that amount, or must be able
to show long-term financial responsibility and stability as measured
by stringent financial tests.

None of these measures address liability for damages to people
or the environment after facility closure. This problem is addressed in the United States by the new Post-Closure Liability Trust
Fund set up in 1980, under the "Comprehensive Environmental Response,
Compensation, and Liability Act".[19] The Fund is to be built up to

a level of \$200 million by a tax of \$2.13 per dry ton of hazardous waste disposed, paid by every hazardous waste disposal facility owner or operator to the U.S. Treasury. If ground-water monitoring at a permitted hazardous waste disposal facility for five years after closure shows no substantial likelihood of risk to public health, the facility owner or operator can transfer subsequent liability for damages to the national fund.

In summary, it can be seen that mechanisms for long-term care are being developed as essential elements in a comprehensive system for hazardous waste disposal. Very few NATO countries have experience with the licensing of hazardous waste landfills, and the United States is the only country currently requiring financial arrangements to assure long-term care.

ACKNOWLEDGEMENTS

The author wishes to acknowledge the particular contributions of Mr. Walter W. Kovalick of the U.S. Environmental Protection Agency and Mr. Peter Mazerolle of Environment Canada to the project reported on in this paper. Mr. Kovalick led the project and was the principal author of the final report. Mr. Mazerolle wrote the report sections on permit processes, labeling, and facility safety.

REFERENCES

1. North Atlantic Treaty Organization/Committee on the Challenges of Modern Society, "Recommended Procedures for Hazardous Waste Management," Report No. 62 (1977).

2. U.S. Code of Federal Regulations, Title 40, Part 261, Identification and Listing of Hazardous Wastes, Federal Register pp. 33084-183 (May 19, 1980).

3. U.S. Environmental Protection Agency, "Test Methods for Evaluating Solid Waste," SW-846, U.S. EPA, Office of Solid Waste, Washington, D.C. 20460 (1980, supplements August 1980 and July 1981).

4. American Society for Testing and Materials, "ASTM Standard D270-23," ASTM, 1916 Race Street, Philadelphia, Pennsylvania 19103.

5. American Society for Testing and Materials, "ASTM Standard D1452-19," Annual ASTM Standards, Part 19, ASTM, 1961 Race Street, Philadelphia, Pennsylvania 19103 (1974).

6. Environment Canada, "Procedures for the Analysis of Landfill Leachate," Appended Seminar Proceeding Report, EPS-3-SW-75-9, Environment Canada, Ottawa, Ontario, Canada (1975).

7. M.J. Taras, A.E. Greenberg, R.D. Hoak, and M.C. Rand, Eds., "Standard Methods for the Examination of Water and Waste-Water," 13th ed., American Public Health Association,

American Water Works Association, and Water Pollution
Control Federation, 1015 Eighteenth Street, N.W.,
Washington, D.C. 20036 (1971).

8. U.S. Environmental Protection Agency, "Manual of Methods for
Chemical Analysis of Water and Wastes," EPA-625-1-6-74-
003, U.S. EPA, Office of Technology Transfer, Washington,
D.C. 20460, (1974).

9. Hansjorg Seng, Mull und Abfall: Fachzeitschrift fur Benhand-
lung und Beseitigung von Abfallen, pp. 148-157, in:
"Standortbeurteilung bei Deponien [Evaluation of Landfill
Sites]" (May 1974).

10. D.A. Gray, J.D. Mather, and I.B. Harrison, "Review of Groundwater Pollution from Waste Disposal Sites in England and
Wales, with Provisional Guidelines for Future Site Selection," Hydrogeological Department, Institute of Geological
Sciences, London (1974).

11. California State Water Resources Control Board, "Waste Discharge Requirements for Waste Disposal to Land--Disposal
Site Design and Operation" (March 1975).

12. R. Piskin, Suitability of Landfills for Disposal of Hazardous
Wastes in Illinois, reprinted from: Waste Age, 7(7):
42-52 (July 1976).

13. U.S. Environmental Protection Agency, "Hazardous Waste Landfill
Site Screening Criteria--Guidance Document," U.S. EPA,
Office of Solid Waste, Washington, D.C. (1976).

14. U.S. Code of Federal Regulations, Title 40, Part 264, Subpart
B--General Facility Standards, Section 264.18--Location
Standards, Federal Register, pp 2848-49 (January 12,
1981).

15. Ministere de la Qualite de la Vie, "Direction de la Prevention
des Pollutions et Nuisances, Ordures Menageres--L'implan-
tation d'un Centre de Traitment," SEMA Marketing et
Modeles de Decision, Paris (March 1976).

16. U.S. Environmental Protection Agency, "Interim Policy Report
on Public Acceptance of Hazardous Waste Management Facilities, " U.S. EPA, Washington, D.C. (April 1976).

17. U.S. Environmental Protection Agency, "Siting of Hazardous
Waste Management Facilities and Public Opposition,"
Report SW-809, U.S. EPA, Office of Solid Waste, Washing-
ton, D.C. (November 1979).

18. U.S. Code of Federal Regulations, Title 40, Part 264, Subpart
G--Closure and Post-Closure and Subpart H--Financial
Requirements, Federal Register, pp 2849-66 (January 12,
1981).

19. Public Law 96-510, 42 USC 9601, "Comprehensive Environmental
Response, Compensation, and Liability Act of 1980,"
(December 11, 1980).

TRANSPORTATION REGULATIONS OF HAZARDOUS WASTE;

U.S.A. AND INTERNATIONAL DEVELOPMENTS

Alan L. Roberts

U.S. Department of Transportation
Washington, D.C., United States

I appreciate the opportunity that has been provided me today to discuss the transportation of hazardous waste. When I was asked to make this presentation, I was not very familiar with the CCMS pilot study on the disposal of hazardous waste, and, to my knowledge, the July 1977 report dealing with transportation has not been forwarded to the United Nations Committee of Experts on the Transport of Dangerous Goods for their consideration. Since preparation of the report, a number of actions, which I would like to discuss today, have been taken concerning the transportation of hazardous waste.

Before I discuss the international aspects of hazardous waste transportation, I will provide a brief description of the United States regulatory program conducted by our Bureau. The Office of Hazardous Materials Regulation is one of the principal offices of the Materials Transportation Bureau, which is an element of the Department of Transportation's Research and Special Programs Administration. The primary mission of my office is development of regulations pertaining to safe transportation of hazardous materials, the issuance of exemptions from certain aspects of the regulations, and to render interpretations of the regulations to all interested parties. The office is divided into three Divisions: the Standards Division, the Exemptions and Approvals Division, and the Technical Division. Reporting to me in my immediate office is a full-time International Standards Coordinator.

Prior to 1975, responsibility for administration of the hazardous materials regulatory program was assigned to the individual administrations in the Department of Transportation. The Coast Guard was responsible for hazardous materials transported by vessel, the Federal Aviation Administration for carriage aboard aircraft, the Federal Railroad Administration for carriage by rail, and the Federal Highway Administration for carriage by motor vehicle. Upon passage of the Hazardous Materials Transportation Act in 1975, the Secretary of Transportation was authorized to merge the regulatory authority for hazardous materials transportation (except shipments in bulk tanks by vessel) into a single organizational entity of the Department, the Materials Transportation Bureau. This means that my office presently is responsible for the development and issuance of all proposed regulations and exemptions pertaining to the transportation of hazardous materials by air, highway, rail, and water. The Director of the Materials Transportation Bureau issues all final regulations. This combined responsibility enables us to seek greater harmony in the intermodal movement of hazardous materials in both domestic and international commerce.

Prior to the 1970's, the focus of our hazardous materials regulatory program was limited to those materials which posed substantial hazards to persons or property in the immediate or direct sense, sometimes referred to as acute exposures. During the 1970's, there were a number of legislative actions in the United States which were to bring about the significant change in the manner in which we conduct our regulatory program. The Occupational Safety and Health Act was designed to assure workers a safe and healthy work envirnoment. An example of an action taken by our Bureau as a result of this legislation is our regulation pertaining to the transportation of asbestos to reduce transportation worker exposure to these materials while they moved in commerce. The second legislative action was the Clean Water Act which led to our promulgation of a list of hazardous substances that are subject to immediate discharge notification requirements and other transportation controls even though they may not necessarily pose acute hazards to people who may come in contact with them during transportation. The third was passage of the Hazardous Materials Transportation Act in 1975 which broadened and consolidated our authority to regulate the transportation of hazardous materials in commerce. And last was passage of the Resource Conservation and Recovery Act in 1976, which is most germane to the topic of consideration at this conference.

Section 3003 of the Resource Conservation Recovery Act (RCRA) requires the Administrator of the Enviornmental Protection Agency (EPA), after consultation with the Secretary of Transportation, to promulgate regulations establishing standards applicable to transportation of hazardous wastes. RCRA requires that the standards issued by the Administrator allow transportation of hazardous waste only if properly labeled and only to permitted treatment storage or disposal facilities designated by the generator of the waste. This designation is required on a manifest to be prescribed by the Administrator. Included also is a requirement that the regulations issued by EPA be consistent with the requirements of the Department of Transportation under the Hazardous Materials Transportation Act. As required, EPA approached us in 1977 to initiate coordination of the new program to deal with hazardous wastes.

In dealing with EPA on hazardous waste matters, we were guided by the following four principles:

O   The existing DOT hazardous materials transportation regulatory scheme is an established and fully operational base that could be adjusted to address risks not then addressed by DOT's regulations.

O   Rather than set up a new regulatory scheme which would add new and different requirements, we preferred to build onto the existing DOT regulatory scheme as a means of minimizing costs to both the government and the private sector.

O   In determining whether the hazardous materials regulatory scheme should be extended to materials not presently covered, DOT deferred to the judgment of EPA in identifying health and environmental risks.

O   In determining how materials identified by EPA as posing a risk are to be contained and their risks communicated while in transportation, EPA should defer to DOT experience and judgment.

By following these four principles we were endeavoring to implement new regulations to control the transportation of hazardous wastes while not imposing a significant number of new regulations upon businesses that have been accustomed to following our hazardous materials transportation regulations for many years.

Our hazardous waste transportation regulations were issued in their final form on May 22, 1980, and became mandatory on November 20, 1980, one day following the effective date of EPA's hazardous waste regulations, including those pertaining to generation storage and disposal of wastes as well as transportation.   A significant action yet to be taken is a proposal for a uniform hazardous waste manifest.   During the past several years, more than 20 State governments have adopted their own hazardous waste manifest forms.   This has created difficulties for generators and transporters of hazardous wastes in the United States.   We have been working with EPA, industry organizations and State officials to resolve this difficult problem and believe that a new regulation adopting a uniform national hazardous waste manifest will be issued within the next six months.   When the new manifest is adopted, all other manifests employed in the United States will be inconsistent with the Federal requirement for use of the uniform waste manifest and, therefore, preempted.   A draft copy of the manifest that has been developed to date is shown on the following pages.

One of the responsibilities of my office is coordination of international activities pertaining to the transportation of hazardous materials, referred to as dangerous goods by international organizations.   The principal organization in which we invest much of our time and effort is the United Nations Committee of Experts on the Transport of Dangerous Goods.   I had the honor and privilege of being elected chairman of the Committee last December.

The products of the Committee's activities are recommendations for the safe transport of dangerous goods.   The recommendations are forwarded to the Economic and Social Council of the United Nations every two years for adoption.   The Economic and Social Council then forwards the recommendations to international bodies concerned with the transport of dangerous goods with a recommendation that they be considered for adoption by those bodies.   For many years now, the Inter-Governmental Maritime Consultative Organization (IMCO) has given full consideration to the UN recommendations in the development of its standards for the safe transport of dangerous goods by sea. More recently, the Dangerous Goods Panel of the International Civil Aviation Organization (ICAO) has adopted new standards which will shortly become mandatory under the Chicago Convention.   The ICAO standards are based primarily on the recommendations of the UN Committee.

UNIFORM HAZARDOUS WASTE MANIFEST				GSA NO. XXX FORM APPROVED OMB NO. XXX			

(Please print or type with ELITE type (12 characters per inch).

TO BE FILLED IN BY THE GENERATOR

GENERATOR NAME AND MAILING ADDRESS

MANIFEST DOCUMENT NUMBER
EPA I.D. NUMBER

AREA CODE/PHONE NUMBER

TRANSPORTER #1 — EPA I.D. NUMBER

TRANSPORTER #2 — EPA I.D. NUMBER

TREATMENT, STORAGE, OR DISPOSAL (TSD) FACILITY — EPA I.D. NUMBER

AREA CODE/PHONE NUMBER

PROPER U.S. D.O.T. SHIPPING NAME AND HAZARD CLASS	UN/NA NUMBER	TOTAL QUANTITY	UNIT WT/VOL	CONTIANER NO |TYPE	WASTE NO. (OPTIONAL)

DRAFT

SPECIAL HANDLING INSTRUCTIONS . . .

THIS IS TO CERTIFY THAT THE ABOVE NAMED MATERIALS ARE PROPERLY CLASSIFIED, DESCRIBED, PACKAGED, MARKED AND LABELED AND ARE IN PROPER CONDITION FOR TRANSPORTATION ACCORDING TO THE APPLICABLE REGULATIONS OF THE DEPARTMENT OF TRANSPORTATION AND THE EPA

DATE SHIPPED
MO DAY YR

SIGNATURE AND FULL NAME PRINTED OR TYPED

☐ CHECK IF CONTINUATION SHEET IS USED. NUMBER OF CONTINUATION SHEETS _____

TO BE FILLED IN BY TRANSPORTER

TRANSPORTER 1 ACKNOWLEDGEMENT OF RECEIPT OF ABOVE MATERIALS

DATE REC'D & ACCEPTED
MO DAY YR

SIGNATURE AND FULL NAME PRINTED OR TYPED

TRANSPORTER 2 ACKNOWLEDGEMENT OF RECEIPT OF ABOVE MATERIALS

DATE REC'D & ACCEPTED
MO DAY YR

SIGNATURE AND FULL NAME PRINTED OR TYPED

TO BE FILLED IN BY TSDF

DISCREPANCY INDICATION SPACE

FACILITY OWNER OR OPERATOR: CERTIFICATION OF RECEIPT OF HAZARDOUS MATERIAL COVERED BY THIS MANIFEST EXCEPT AS NOTED IN THE DISCREPANCY INDICATION SPACE ABOVE

DATE REC'D & ACCEPTED
MO DAY YR

SIGNATURE AND FULL NAME PRINTED OR TYPED

UNIFORM HAZARDOUS WASTE MANIFEST					GSA NO. XXX FORM APPROVED OMB NO. XXX		

(Please print or type with ELITE type (12 characters per inch).

CONTINUATION SHEET

THIS IS CONTINUATION SHEET____ OF____

MANIFEST DOCUMENT NUMBER
EPA I.D. NUMBER

TRANSPORTER #   EPA I.D. NUMBER

TRANSPORTER #   EPA I.D. NUMBER

PROPER U.S. D.O.T. SHIPPING NAME AND HAZARD CLASS	UN/NA NUMBER	TOTAL QUANTITY	UNIT WT/VOL	CONTAINER NO	TYPE	WASTE NO. (OPTIONAL)

DRAFT

TO BE FILLED IN BY THE GENERATOR

TRANSPORTER ACKNOWLEDGEMENT OF RECEIPT OF ABOVE MATERIALS   DATE REC'D & ACCEPTED
                                                            MO   DAY   YR
SIGNATURE AND FULL NAME PRINTED OR TYPED

TRANSPORTER ACKNOWLEDGEMENT OF RECEIPT OF ABOVE MATERIALS   DATE REC'D & ACCEPTED
                                                            MO   DAY   YR
SIGNATURE AND FULL NAME PRINTED OR TYPED

TO BE FILLED IN BY TRANSPORTER

Within the last several years, two regulatory bodies in Europe, identified in short as RID (rail transport) and ADR (highway transport), have given serious consideration to the UN recommendations in revisions to their standards. As you can see by the numbers of organizations I have identified, participation in these activities is a full-time occupation. Several years ago, we established a full-time International Standards Coordinator position in my office to deal with these matters both at home and at international meetings.

Since 1974, our delegation to the UN Committee has attempted to persuade the heads of other delegations to give greater consideration to environmental matters relative to the transportation of goods in world commerce. Early in this period, the UN Committee did decide to list polychlorinated biphenyls as dangerous goods; however, our efforts to get many other environmentally hazardous materials listed has met with little success. A principal effort was initiated by a paper submitted by the representative from Canada which contained a listing of all the hazardous substances we adopted in the United States last year. On two occasions, the matter was brought to a vote and the results were the same. The United States, Canada, the USSR, and Japan voted in favor of adoption and the Federal Republic of Germany, the United Kingdom, Italy, and France voted against adoption; therefore, the proposal was not adopted on either occasion. Subsequently, the United States proposed that the word "waste" precede any UN description on a shipping document when a material was being transported for disposal or processed for disposal. This proposal met with little success during several sessions of the UN Committee's subsidiary bodies; however, at its December 1980 meeting, the Committee voted to adopt the proposal and it now appears as a recommendation in Chapter 13 of the Committee's recommendations.

I am reasonably confident that the waste recommendation will be adopted by IMCO as a mandatory requirement and hopefully by the other international bodies in the near future. With this simple amendment to the UN recommendations, and its adoption by the operative bodies, it will be mandatory that shipping papers contain a disclosure that a material is a hazardous waste being transported for disposal. This is important, since the entry will be one means of mandatory disclosure of the presence of a hazardous waste in a shipment that should be accessible to governmental authorities at points of entry into various nations, including developing nations.

I mentioned earlier that we have one major action to take in the United States concerning hazardous waste, and that pertains to the adoption of a uniform national hazardous waste manifest. I would also suggest that similar consideration be given to a standard hazardous waste manifest format for worldwide commerce. If our past record at the Committee is any indication, I visualize that it will take many years before such an international requirement is adopted. In the meantime, we must rely on the description specified on shipping documents as implemented by international and national regulatory agencies.

We must also continue our attempts to have environmentally hazardous substances listed in the United Nations recommendations. I think the concern here is over the added burdens to commerce that would follow. I personally do not believe this has to happen. For example, in the United States we have added quite a number of materials to our list of hazardous materials placing emphasis on communication (package marking and shipping papers) and discharge reporting. We have not added significant burdens to shippers and carriers relative to labeling, placarding, and other special controls that apply to materials that pose an acute hazard during transportation. The UN Committee can do likewise by adding to its Class 9 many of the materials that can cause environmental harm, thereby imposing very limited requirements relative to their transportation. I would encourage each of you coming from countries having a voting representation at the UN Committee to encourage your degelations to support such a proposition, and possibly an international hazardous waste manifest. For your information, the countries participating in the UN Committee that have a vote are Japan, Canada, United States, United Kingdom, France, Federal Republic of Germany, Italy, Poland, Norway, USSR, Thailand, Iraq, and Iran, although the latter three have not been active at sessions of the Committee during the last several years.

In summary, there are established systems of standards and requirements in effect for the transport of dangerous goods in worldwide commerce. They should be expanded as necessary to include the transportation of enviornmentally dangerous goods - including hazardous wastes.

STATUS OF CHEMICAL, PHYSICAL AND BIOLOGICAL TREATMENT PROCESSES

IN HAZARDOUS WASTE MANAGEMENT

Eugene Crumpler

Hazardous & Industrial Waste Division
U. S. Environmental Protection Agency
Washington, D.C. 20460

INTRODUCTION

In 1978, the NATO/CCMS Phase II Pilot Study on Disposal of
Hazardous Waste started a project to develop information on the
state-of-the-art of chemical, physical and biological (C/P/B)
treatment processes as applied to hazardous wastes. This effort
was directed at making member countries and waste generators
aware of the alternatives to land disposal and incineration for
management of industrial wastes. This paper summarizes the
results of that effort which produced the following major outputs:

- ° A summary of waste management policy which effects C/P/B
  treatment applications in each country.

- ° A waste/treatment process matrix to quickly identify
  potential matches of waste and C/P/B processes.

- ° A discussion of the future potential for recovery of
  hazardous wastes by C/P/B processes, and

- ° A summary of centralized treatment of wastes and current
  research and development efforts in the NATO countries.

SCOPE AND METHODOLOGY OF PROJECT

The subproject took the form of an information gathering
effort. The United States first sent a questionnaire to each

participant, requesting information on C/P/B treatment processes
either being used full-scale or under development in their country.
The original request asked for a reply identifying the types of
waste material treated by unit process. After receipt of the
initial response the United States followed up with a second
questionnaire asking for information by wastes.

Written contributions were received from the Federal Republic
of Germany, France, the Netherlands, the United Kingdom and the
United States. The state-of-the-art report titled, Physical,
Chemical, and Biological Treatment Techniques for Industrial Waste[1]
completed by the United States Environmental Protection Agency in
1977 also was used as a basis for the project (A copy of the executive summary of that report is appendixed to the final NATO/CCMS
project report).[2]

The final project report does not contain a comprehensive
list of all applications of C/P/B processes to wastes. The possible
combinations of processes and wastes is too great to adequately
cover in a single effort. This field is expanding rapidly due to
high interest in the developed countries. The report was designed
instead to serve as a source of ideas for those who are dealing
with the problem of hazardous waste. A waste/process matrix is
the core of the report.

HOW INDIVIDUAL HAZARDOUS WASTE POLICY EFFECTS EACH COUNTRIES APPROACH TO C/P/B TREATMENT

During the early stages of the project, it was obvious that
each country's laws and national and regional policies impact the
extent that C/P/B treatment is or will be used for waste management.
Each country is approaching its hazardous waste program from
different points of view. Examples of these policies are as follows:

Federal Republic of Germany

The Federal Republic's 1972 Waste Disposal Act and 1975 Waste
Management Program govern policy. Selection of treatment and
disposal methods must consider costs, risks, and benefits for a
particular waste. In some cases, certain wastes are assigned to a
particular treatment/disposal method in the facility license.
Where appropriate, pretreatment methods may be specified.

The disposal policies of each Federal State affect the
treatment/disposal method used. A State operating a state-wide
company with a broad range of treatment methods tends to exclude
co-disposal of many types of industrial wastes at municipal

landfills. States with limited facilities tend to allow co-disposal
with the waste generator handling pretreatment, if needed.

The Netherlands

The 1976 Chemical Waste Act prohibits land disposal except
for special or urgent conditions. Thus all forms of treatment
are preferred. Reduced waste generation, recycling, treatment,
and burning as a fuel are alternatives being actively pursued.
All chemical wastes are extensively controlled from generation
to final disposal.

United Kingdom

Two major laws govern waste disposal in the United Kingdom;
the 1972 Deposit of Poisonous Waste Act and the 1974 Control of
Pollution Act. The Deposit of Poisonous Waste Act creates penalities
for deposition of hazardous waste on land if it might damage
humans, or animals, or pollute water supplies. It also requires
notification of local official of waste activities in their
jurisdictions. The Control of Pollution Act requires licensing
of treatment and disposal sites.

The United Kingdom's policy on hazardous waste
management is as follows:

> ° The generation of wastes should be minimized, and
>   wastes should be reused and recycled, if economically
>   and technically feasible.
>
> ° All waste must be managed at licensed facilities.
>
> ° Economic considerations will be of primary importance
>   in determining which licensed alternatives are selected
>   for waste management.

United States

Hazardous waste policy in the U.S. is being developed under
four major acts:

> ° The Federal Insecticide, Fungicide, and Rodenticide
>   Act of 1972 (FIFRA)
>
> ° The Safe Drinking Water Act of 1974
>
> ° The Toxic Substances Control Act of 1976 (TSCA)

° The Resource Conservation and Recovery Act of 1976
(RCRA)

The specific mandate to control hazardous waste is contained
in Subtitle C of RCRA. The act requires that:

° Hazardous waste be identified

° A system to track waste from generation to final disposal
be created

° treatment, storage, and disposal facilities be permitted
based on minimum national standards

The intent of RCRA is that hazardous waste regulatory control
be handled by the individual States. EPA will determine if each
State's program meets minimum Federal standards. The Federal EPA
will manage the program in any State that chooses not to accept
the responsibiility or does not meet minimum Federal guidelines.

Land disposal in the United States, which has been the primary
disposal method in the past, will become much more costly as the
new standards are applied. Public awareness of the dangers of
land disposal will make C/P/B treatment and incineration more
attractive alternatives. Also, EPA is encouraging these technologies
through research, demonstration projects, and regulatory policy.

Other NATO countries are developing or have in place similar
policies, for hazardous industrial wastes. The policies presented
above are consider representative.

PROCESS/WASTES MATRIX

A matrix of C/P/B treatment processes and types of hazardous
industrial wastes was developed to compile the contributions of
the member countries and provide a fast reference device (Table 1).
The following information is included in the matrix:

° each process waste/combination has been classified as
1, full-scale use; 2, moderate use; 3, research stage

° the cost of the treatment processes has been generally
classified as being low, medium, or high to provide
guidance in evaluating the waste/treatment combinations

POTENTIAL RECOVERY OF HAZARDOUS WASTES

The potential for greatly increased recycling exists in all NATO countries. In the past, disposal of wastes has been the lowest cost management method. Implementation of national policies on waste disposal will change the economics of these practices.

Federal Republic of Germany

Waste oils and solvents are the only hazardous wastes presently being extensively recycled in the Federal Republic. About 100 recycling plants are currently in operation. The cost of virgin raw materials puts a ceiling on recycling costs. Much effort in the Federal Republic is aimed at developing technologies that produce little or no waste.

The Netherlands

Efforts in The Netherlands are primarily directed to limiting waste generation. Major recycling efforts are directed to metal-containing wastes, such as catalysts, plating wastes, copper sludges, pickling baths and photographic wastes. Storage for 5 to 10 years is a part of the overall waste management plan until research and development efforts can be turned into commercial processes.

United Kingdom

The Chemical Recovery Association, a nonprofit organization, is the focus of resource recovery in the United Kingdom. Several plants are being brought on stream to recover metal fininshing wastes.

United States

The U.S. is working under the RCRA to develop the following strategy for resource recovery:

- ° reduce generation of waste by process modifications

- ° recover materials and energy from waste to maximize resource recovery

- ° for those wastes that cannot be eliminated or recovered, insure that management methods protect human health and the environment.

Table 1.   Application Matrix of Chemical, Physical, and Biological Treatment Process to Wastes.

Process	Relative Cost Factor	Petrol. Refin.	Petrochemicals	Pharmaceutical	Munitions	Paint&Ink Ind.	Painting Waste	PVC Production	Leather Tan.	Food Process.	Metal Fin. Liq.	M.F. Sludges	Pickle Liquor	Metallurg. Slag	Waste Oil	Chlor-Alkali	Carbonyls	Fluoride	Organic Acids	Inorganic Acid	Mercury	Cadmium	PNA's	Amines	Aromatics	Aliphatics	Ethers	Phthalates	Phenolics	Chlor. HC's	Cyanide	Sulfides	Pesticides	PCB's	PCP	2378-Dioxin
carbon adsorption	Low	1	3		1	3	3	2			1				1	1					2	3	2	2	2	2	2	2	1	1		3	1	1	2	2
oil sep techniques	Low	1	2	2	3	3		3		3				3	1			2	1	2				3	3				3	3	1		3			
wet air oxidation	High		3	2							1				1	1						2		3							1					
neutralization	Low	3		1	3	2	2		2	1	1	1	2	2	1	1		2	1	2					1	1	1	1	3	2	1	2	2			
chem fixation/encapsulation	Med.					2					1	2	2	1		1						2				1	1			1	1	1	3			
filtration	Low	3		1		3				1	1	3			1							1	2		1				3	2	1	1				
ox-reduction	Med.	1	2	1		2	2	2	2	1	1				1	1			2			3	1	2	1	1	1	1	1	2	1	2	3	3	2	3
chem-precip/settling	Low	1				2				1	1	3				1					2	1							1	1	3	1				
evaporation	Low	1		1		1	1	3			1					1			2							1	1		1	1		2				
acid treatment	Med.	1		1	3	3	2	2			1					1			2					1	1	1	1		1	1		1	2			
distillation	Low	2	3	1		1	1			1	3	3			2	3		3		2				3	3	3	2	3		1	3			3		
biotreatment	Med.	1				2					3						3						1	1	3	3			3	1		1	3	3		
calcination	High				3						3	3																	3	3	3	1	3	3	3	3
flotation	Low	1				1	1			1																						2				
hydrolysis	Low	2		1		2		2							2	3			2								2			3	3	1	2			
liq-liq extraction	Low to High	1									3			2	3		3		2	2			3	3	3	3	2	3	2	1		2	3	3	2	3
ozonation	Med.			2													3						3		3				2	2	2	2	3	3		3
chem dechlorination	High										2	2			2														3	3	3		2	3		
UV dechlorination	High																2												3	3			3	3	3	
catalysis	High													2																			2			
electrodialysis	Low						3				2																									
Reverse Osmosis	Med.	2			2					2	2	2	2		2				2	2		2	2		3	3		2	3	2		2	2	2		
dissolution	Low to Med.					3				2	1		1			1			1			2							3	3			3			
ion exchange	High	2			2					3	3	3	2	2								3	3					3	3	3	2		3	2	2	3
microwave	High																																			
chlorinalysis	High																													3	3			3		3	2
electrolysis	Med.	2					2			3	3			2	2		2	2				3						2	3	3	3	2	2		2	2
ultrafiltration	Low to High	1								2	3	3	3																3	3						
resin adsorption	Low	1									3	3				1						3	2	2	2	2	2	2	3	3	3		2	2		

Use Codes

1.   Full Scale Common Use
2.   Moderate Application
3.   Research Stage

Many waste exchanges have become active in the U.S., although only a small percentage of wastes are currently suitable for exchange. Government action to eliminate disincentives for use of recovered material is proceeding.

WASTE TREATMENT CENTERS IN NATO COUNTRIES

The concept of centralized waste treatment facilities is relatively new, and is being pursued in many NATO countries.

Denmark

In 1971 the Danish municipalities established a national corporation to handle hazardous waste. Kommunekemi is the result of the effort. The following actions were required to establish the corporation:

° A law was passed covering disposal of dangerous wastes.

° Waste estimates were made.

° An organization to collect and transport the waste was developed.

° Funds for construction and operation were established.

° Construction of collection stations and the centralized treatment and disposal plant was started. The original capacity was designed for a five year period.

The chemical waste regulations were implemented in 1976. All industry was required either to deliver their wastes to one of 20 collection stations or show it was effectively managing the wastes on site. From the central stations wastes are shipped by rail to the central plant in Nyborg.

The Nyborg facility currently has the following operations:

° storage and laboratory facilities

° waste oil treatment plant

° inorganic chemical waste treatment plant (C/P/B)

° two incinerators

° controlled landfill.

Future facilities planned for 1982 include:

    ° a new rotary kiln incinerator

    ° a larger oil treatment plant.

Federal Republic of Germany

The major integrated TSD facilities in the Federal Republic are located in the State of Bavaria. These are operated by (1) Zweokverband Sondermullplatze Mittelfranken located in Schwabach since 1968 and (2) Gesellshaft for Sondemullbeseitigung in Bayern, (GSB) located near Munich since 1971.

These centers receive special wastes which are treated by:

° Neutralization and detoxification

° Dewatering

° Clarification of wastewater prior to discharge

° Incineration

° Disposal in secure landfills.

The plants are served by seven special waste collection centers in Bavaria. The centers have testing facilities, large temporary storage, some pretreatment, and oil separation systems. By pretreating, the received material is reduced to less than 10 percent of the original volume which is then transported to the two facilities.

The industries transport their wastes to the collection centers either by their own trucks or hired trucks. Wastes are moved by container or tank car, from the collection centers to the two disposal plants. In 1979 GSB received and disposed of 190,000 metric tons of waste.

Other States of the Federal Republic do not have such integrated treatment and disposal facilities. However 20 smaller treatment facilities are distributed all over the country. Those centers offer detoxification (cyanide and chrome), neutralization, oil emulsion separation, and sludge dewatering. Some also operate their own secure landfills.

France

France currently has four integrated waste management centers.
Saint Vuilbas Center - The Saint Vuilbas center has been operated
by Plafora, SA. since 1975. It contains four principal processes:

   ° a rotary kiln incinerator with a wet scrubber
     (20,000 mt/yr)

   ° a physical-chemical detoxification process for chemical
     removal, neutralization, catalytic and electrolytic
     oxidation, and oil separation (18,000 mt/yr)

   ° an ion-exchange resin regeneration unit

   ° a sludge dehydration unit (5,000 mt/yr).

Mitry Compans Center - The Mitry Compans Center has been operated
by GEREP-CA since 1977. The center consists of two principal
units:

   ° rotary kiln and liquid injection incinerators served
     by a common wet scrubber system (18,000 mt/yr)

   ° A fixation process (SOLIROC) (5,000 mt/yr).

Limay Center - This center was started in 1975 by SARP Industries.
It contains four principal units:

   ° Liquid injection incinerator (15,000 mt/yr)

   ° Treatment systems using neutralization, precipitation,
     chromium removal, and cyanide destruction by catalytic
     oxidation.

   ° copper removal process for printed circuit wastes
     (3,000 mt/yr)

   ° a fixation process (Chemfix) (100,000 m^3/yr).

Hombourg Center - PEC Engineering has operated the Hombourg
Center since 1974. The center contains:

  ° a treatment unit consisting of neutralization, precipitation,
    chromium removal (SO_2), and cyanide destruction by
    catalytic oxidation (27,000 mt/yr)

° a fixation process (Petrifix) (30,000 mt/yr)

° a regeneration unit for ion exchange resin cartridges (1,000/year).

United Kingdom

Several privately owned waste treatment centers currently operate in the United Kingdom. These centers offer processing of inorganic wastes and incineration of organics. Products from these treatments are sometimes reused; otherwise they are sent to licensed landfills.

In addition to the large integrated facilities, there are also several small waste treatment companies in the United Kingdom. Typically, neutralization of acids, reduction of hexavalent chromium, and precipitation of metal hydroxides are carried out. Several companies recover waste oils. A central ion-exchange regeneration facility is operating in a region where metal finishing wastes are produced. One company mixes latex wastes, heavy oils, and paint wastes with coal fines to make a fuel for firing a local power station. A copper refiner takes in copperbearing wastes. Several companies take in ammonium and phosphate wastes in order to produce nitrogen/ phosphorous/ potassium fertilizers. There are, of course, many in-house waste treatment plants utilizing various processes, but relevant information has been difficult to obtain.

Without doubt, a great amount of waste is being treated in the United Kingdom. The Department of the Environment, through its Waste Management Paper Series, is endeavoring to further this by encouraging centralized facilities, particularly in the case of metal-finishing and solvent wastes.

United States

There are a number of treatment and disposal facilities operating in the United States which accept a variety of hazardous waste materials. The majority of facilities are owned and operated by private industry. The Federal regulations issued under RCRA will have significant impact on the activities of these treatment and disposal facilities. Some facilities may have to improve their operations to meet the regulations. A number of companies are interested in the hazardous waste management industry, but will not begin serious planning until the final regulations are issued.

It is estimated that there are at least 313 off-site facilities in the United States. Disposal technologies include incineration (rotary kiln, liquid injection, fluidized bed, and other methods) and land disposal (secure landfill, sanitary landfill, soil incorporation/ landfarming, encapsulation, lagooning, and other methods). Of the 313 facilities, at least 44 practice some form of incineration, and 117 practice some form of land disposal. At least 123 facilities are involved in collection/ hauling, 89 in processing/treatment, and 116 in recycling/ reclamation.

Typical of the central off-site facilities expected in the future is one being designed by a major waste management corporation for a highly industrialized region in Southern Louisiana. This facility is to be operated by a private corporation, but the funding is to be through the sale of State guaranteed tax exempt bonds.

The processes planned for this facility include: acid distillation, toluene distillation, solvent extraction, rotary kiln incineration with heat recovery, chemical precipitation and neutralization, carbon absorption, chemical fixation, activated sludge, screening, clarification, dewatering, treatment of effluent, and secure landfill.

RESEARCH AND DEVELOPMENT OF EFFORTS

Principal research efforts on C/P/B Treatment are listed below by country.

The Netherlands

   ° Immobilization techniques for metal wastes

   ° Metal extraction and leaching processes

United States

   The U.S. EPA is conducting research in the following areas:

      ° leachate treatment by C/P/B processes

      ° biological treatment of recycled sanitary landfill leachate

      ° studies of C/P/B processes for pretreatment of inorganic wastes discharged to sewers

° carbon treatment of military pesticide wastes

° dechlorination of PCB type compounds using sodium
    polyethylene glycol

° studies of pesticide disposal by wastewater biological
    treatment methods

° development of a PCB dechlorination process using
    sodium napthalide.

° evaluation of a process for the ultraviolet destruction
    of 2,3,7,8-tetrachlorodibenzodioxin.

United Kingdom

The United Kingdom is investigating the microbiological
sequestration of heavy metals, initially cadmium, through the
development of highly selective strains of bacteria; also the
biodegradation of intractable organic wastes, such as PCB, through
the development of specialized bacteria.

(The reader is referred to the complete report[2] for additional
details on the above research and development efforts.)

CONCLUSIONS

(a) Most chemical, physical, and biological treatment processes
     can reduce the volume or hazard of waste. However, a residue
     usually remains that may require further treatment or disposal.
     Hence, chemical, physical, and biological treatment by
     itself is not always a cure-all.

(b) Some conventional chemical, physical, and biological
     treatment processes (such as neutralization) are well
     developed and widely applied to hazardous wastes at both
     on-site and centralized off-site facilities. Other processes
     are specific to particular wastes and are usually applied
     by the waste generator at on-site facilities.

(c) Chemical, physical, and biological treatment processes have
     the best potential to recover resources from wastes. Such
     recovery can play an important role in reducing the quantities
     of waste going to land and ocean disposal.

(d) Technology for resource recovery and recycling of materials
     from hazardous waste is still in its early stages of development,

except for waste oil and waste solvents.

(e)   In most NATO countries, information on chemical, physical,
      and biological treatment of wastes on-site is proprietary
      and therefore is difficult to obtain.

(f)   In many NATO countries, chemical, physical, and biological
      treatment processes are widely used. Restrictions on land
      disposal, specific regulatory requirements for resource
      recovery, and financial incentives will further increase
      resource recovery by chemical, physical, and biological
      treatment.

(g)   The recent increase in construction and operation of centralized
      treatment facilities in European NATO countries is expected
      to be repeated in Canada and the United States following
      implementation of regulatory programs. Much activity is
      anticipated over the next 5 to 10 years.

(h)   In many NATO countries, centralized chemical, physical, and
      biological facilities are for the most part privately owned.
      However, in a few cases, centralized facilities have been
      partially or wholly underwritten by regional or local
      authorities.

(i)   Research and development of chemical, physical, and biological
      treatment is being actively pursued by national governments,
      industry, and other organizations in many NATO countries.

RECOMMENDATIONS

(a)   In evaluating hazardous waste management alternatives,
      resource recovery by chemical, physical, and biological
      processes should be encouraged before thermal treatment, or
      land or ocean disposal.

(b)   Use of chemical, physical, and biological treatment processes
      for hazardous waste should be encouraged as a means to
      improve environmental protection.

(c)   The research and development efforts of the member NATO
      countries, particularly in resource recovery from wastes
      and treatment of chlorinated organic wastes, should be
      encouraged where these processes are appropriate for
      environmental protection and are expected to be economically
      and technically viable.

(d)   Increased efforts are needed among nations to exchange
      information on hazardous waste treatment methods. Development
      of a uniform hazardous waste nomenclature system would
      facilitate this exchange of information.

(e)   The role of chemical, physical, and biological treatment,
      including both on-site and centralized facility application,
      should be further evaluated in the context of the overall
      hazardous waste management system. The Organization for
      Economic Cooperation and Development (OECD) would be an
      appropriate organization for such an evaluation.

(f)   Many chemical, physical, and biological treatment processes,
      especially those with potential for resource recovery, can
      and should be applied to non-hazardous industrial wastes,
      where technically and economically feasible, to reduce the
      amounts of waste requiring land or ocean disposal.

REFERENCES

1. U.S. Environmental Protection Agency, "Physical, Chemical,
    and Biological Treatment Techniques for Industrial Waste",
    USEPA, Office of Solid Waste, Washington, D.C. (1977)

2. NATO/CCMS Pilot Study on Disposal of Hazardous Wastes,
    "Chemical, Physical, and Biological Treatment of Hazardous
    Wastes in NATO Countries", NATO-CCMS Report No. 119. (1981)

BIBLIOGRAPHY

U.S. Environmental Protection Agency "Alternatives for Hazardous
Waste Management in the Metals Smelting and Refining Industries",
USEPA, Office of Solid Waste, Washington, D.C. (1977)

U.S. Environmental Protection Agency "Alternatives for Hazardous
Waste Management in the Organic Chemical, Pesticides, and Explosives
Industries", USEPA, Office of Solid Waste, Washington, D.C. (1977)

U.S. Environmental Protection Agency "Alternatives for Hazardous
Waste Management in the Petroleum Refining Industry", Washington, D.C.
(1977)

U.S. Environmental Protection Agency "Alternatives for Hazardous
Waste Management in the Inorganic Chemicals Industry", USEPA,
Office of Solid Waste, Washington, D.C. (1977)

CHROMIUM CYCLE

METAL FINISHING WASTES

Etienne Le Roy

National Agency for the recovery
and the disposal of wastes
Angers - France

SUMMARY

Significant savings in the chromium consumption can be
achieved firstly in the metallurgical industry, particularly
by recycling blast furnace dusts, scrap-steels and machining
waste-ends, secondly in the chemical industry,i.e., mainly in
the metal finishing shops, the tanneries, and the recovery of
waste catalysts. In chemical industry, metal finishing alone
is responsible for chromium losses which reach more than 70 %
of the total amount of the chromium consumed in this branch of
activities, and which are hexavalent and highly toxic. These
losses, as well as those of other additives, can be significantly
reduced by applying new technologies which resort to new
methods of ion-exchanges, electrolysis, evaporation and membranes. This control policy combined with captive shop
facilities induces a modification in the standard procedures of
effluent treatment. Recommendations are proposed for a better
control of the metal finishing shops and for an improved treatment of the inevitably produced wastes.

INTRODUCTION

France has proposed to contribute to the first phase of
the study concerning the treatment of hazardous wastes undertook
by the Committee on the Challenges of Modern Society (NATO),
by engaging in a specific study on chromium wastes. This choice
was justified by the gravity and extension of the problem of
chromium pollution and by the urgent need for a determined National policy as regards the treatment of these wastes.

Following this report on the chromium cycle, France has offered the pilot group to participate in the second phase of the study on the treatment of hazardous wastes, by focussing on the problems raised by the metal finishing industry. It has, in fact, proved worthwhile to overlap the treatment approach to hazardous waste, all activities mingled, with the analytic approach of the industrial sectors which produce them.

The scope of the report concluding this project on metal finishing wastes has been significantly increased because of a large contribution of information from the United States, the United Kingdom, the Netherlands, Canada, and the Federal Republic of Germany. This exchange of information makes it possible to emphasize the constraints of this branch of industrial activities and to foresee more acutely the impact of technological progress and cost increases in this field.

CHROMIUM CYCLE

Chromium is used in three industrial branches :

. metallurgy
. chemistry
. refractories

Metallurgical industries alone use 90 % of the National chromium consumption which is of almost 100 000 t. in France.

Chromium waste flows are estimated to be of 10 % of the total consumption in this branch. In these wastes, chromium is generally a non-toxic mineral. Dusts fly to the atmosphere during chromite manipulations and ferrochrome solid wastes are disposed of to land.

Chemical industries use 6 % of the National chromium consumption. Waste flows amount to 50 % of this consumption. In these losses, chromium is generally trivalent when not hexavalent, i.e., particularly toxic.

Refractory manufacturing corresponds to 4 % of the National chromium consumption. The manufacturing produces only few losses which are not well known ; henceforth, this point will not be developed any further. It still has to be mentioned that refractories cannot be used twice and that consequently the chromium which is used is lost.

Metallurgy

The recovery of stainless steel waste pieces and of low-grade chromium scrap steels can be developed, especially those of small size. A few percents of metallurgical chromium could thus be recovered, i.e., about 4,000 T in France, and be added to the scrap steels which are already used in the steel-manufacturing industry.

The machining of stainless steels produces around 25 % of waste-ends. It is estimated that only 2/3 of these waste-ends are recovered. By increasing the quantity of recovered waste-ends especially in small manufactures, a few percents of metallurgical chromium, i.e., about 4,000 T could be saved. These waste-ends could be added to the machining scraps of the semi-manufactured products, that are used in the steel manufacturing industry.

These various recovered losses will allow a proportional reduction of the ferrochrome that are used in the steel manufacturing industry.

As regards the elaboration of ferrochromes, the only parts in which it could be possible to recover some of the chromium losses are the calcination in revolving furnaces and the reduction in arc furnaces. As a matter of fact, in both cases, furnace dusts which are responsible for most of the chromium losses, would have to be recovered with specialized equipment. Chromium recovery also requires the implementation of dust aggregation processes, enabling dust recycling.

Given that the percentage of dust recovery is 80% in revolving furnaces and 70% in arc furnaces, it would be possible to avoid 50% of the chromium losses due to the elaboration of ferrochromes.

The totality of these savings would naturally entail a reduction of the amount of the consumed metallurgical chromites.

Chemistry

Chromites are mainly consumed in the sodium bichromate production. After an eventual transformation into sulfate, alum, chromic oxide, or chromic acid, bichromate is used in numerous industrial activities such as tanneries, metal finishing, catalysts, pigments (lead or zinc chromates). Chromium compounds other than sodium bichromate are used in dyeing, photography, electronics, etc.

The greatest amount of chemical chromium losses results from the sodium bichromate manufacturing. The rather low efficiency of the sodium bichromate preparation reaction accounts for the high percentage of losses induced, which are mainly trivalent chromium losses. But there is little hope for any improvement of the process, and consequently, researches shoud be directed towards a new chemical chromite attacking process.

As for bichromate consumption, the three following industrial activities are responsible for more than half of the chromium losses: metal finishing, tanneries and catalysts.

Catalysts stand for about 10 % of the losses of consumed bichromate. It is estimated that 50 % of the contained chromium could be recovered. But this involves the treatment of significant quantities of catalysts when the recovered chromium would not amount to more than a few tons, since the average catalyst chromium content is of approximately 2 to 3 %.

Tanneries originate about 12 % of consumed bichromate losses. It is thought that 75 % of the chromium contained in the waste baths and waste rinse waters could be recovered. The methods which should be applied are somewhat similar to those recommended for the recovery of chromium in metal finishing.

And as for metal finishing, it is responsible for 35 % of the consumed bichromate losses. Moreover, chromium which is found in waste baths and waste rinse waters is mainly hexavalent and is particularly toxic for man and its environment. 90 % of this waste chromium could possibly be recovered. The subject will be developed in the study on the metal finishing wastes.

METAL FINISHING WASTES

Any action undertaken in order to modify one or several physical, chemical or electrical characteristics of a surface involves a process which will be a treatment of this surface.

These metal finishing operations are required in numerous industrial productions such as automobiles, aeronautics, engineering, etc. A distinction should be made between captive shops which are part of a wider production process and job shops where the chief activity is metal finishing. The latter are fewer and easier to identify than the former. The great number of shops with a very small labour-force or with no wage-earners at all is an essential datum of this activity.

The metal finishing methods that are generally applied, use chemical or electrochemical processes which generate very significant wastes in toxic and costly products.

The sequences of plating operations

The amount of wastes that can be controlled will vary with the processes used to achieve the objectives sought by metal finishing. These processes involve one or the other of the following methods.

Deposition on the surface of the substrate of a different kind of metal possessing the desired characteristics.

The deposition can be achieved by chemical or electrochemical methods. This second process is more common and it is particularly polluting. The baths are often cyanided with a greater metal concentration and a greater susceptibility to pollutants than electroless plating baths. In sequences of plating operations the implementation of methods allowing additive losses to be controlled is always recommended. A decrease in metal, cyanide and chromic acid consumption can thus be obtained at a lower cost than that of purification processes.

Etching or physico-chemical modification of the surface.

In this case, the metal contained in waste baths or in waste rinse waters comes from the substrate. Hence the control procedure which only applies to additive losses, cannot be implemented. The chemical agents are mainly acids and bases with various concentrations and for which a final neutralization is more economical than the control of losses. However, a number of agents, chromic acid among them, should be recovered and recycled. Moreover, the recovery of the metals issued from the substrate, when technically possible,allows their valuation (e.g. electrolysis of copper pickling waste baths).

We are restricting our subject to the sequences of electroplating operations which are the most threatening for our environment. The following metals are the most used : nickel, zinc, copper which represent respectively about 45%, 30%, 8% of the implemented bath capacity; cadmium and tin for all of which losses vary from 5 to 10%, and chromium, about 15% of the bath capacity with losses approaching 70%.

20 to 25 % of these losses are usually concentrated effluents with a content measured in g/l. The remaining are diluted effluents with a content measured in mg/l.

Losses

A summary input/output balance shows that losses proceed
from:

. waste baths
. rinses
. evaporation

Waste baths, losses due to maintenance

Waste baths are concentrated effluents with miscellaneous
salt with contents varying from a few grams to a few hundred grams
per liter. Their total volume, however, remains rather insignificant
In the last few years,improvements have been achieved which
result in the lengthening of the plating bath lifetime or in the
diminution of their concentration. The amount of pollutants
discharged with the waste baths can thus be reduced. These
two loss control procedures are complementary, not in the least
in competition.

Rinses

Rinses are required after most of the plating baths since
pieces emerging from a bath drag out a certain amount of the bath
solution, which, when transferred into the following bath, could
spoil it.

Waste rinse waters are diluted effluents with miscellaneous
salt contents varying from a few milligrams to a few hundred
milligrams per liter, but their total volume is thousands
of times greater than that of waste baths.

Rinsing methods proceed from two different standard procedures :

. running rinses in which the pure water input flow is equal
to the polluted water output flow. In such a bath, contents in
chemical agents are theoretically constant.
. dead or still or recovery rinses in which the chemical agent
content gradually rises to a concentration decided upon the
desired efficiency of the rinsing operation.

Waste rinse waters can be concentrated and eventually
purificated in order to be reintroduced into the plating bath.
In the case of a hot plating bath (40 ° to 70°), waters from
a dead rinse or from a counter-current cascade rinse may be
used to balance the plating bath by compensating the losses
by evaporation.

Evaporation

Vapours issued from the hot plating baths would not be
considered as a source of additive losses. However, in some
cases, with a low cathodic efficiency, e.g., chromium plating,
electroplating process initiates a strong gas release responsible
in turn for the loss of part of the liquid which has the same
characteristics as a plating bath.

Vapours, charged for example with some chromic acid, must
be condensated and recycled into the plating bath.

Control process

All additive loss control processes offer common advantages,
among which:

. raw product savings
. decrease in the volume of the sludges produced by the effluent
  treatment,
. decrease in the reagent consumption in the effluent treatment,
. decrease in rinse water consumption,
. decrease in the size of the effluent treatment facilities,
. decrease in the discharged salinity following effluent
  treatment.

It is to be noticed that the investment required for the
implementation of complementary equipment will often be rapidly
payed off in consideration of the above cited advantages.

The main disadvantage of these processes relies in the
extra floor space required.

Additive loss control requires the optimization of the
straining and rinsing of the pieces that come out of the plating bath. Although these methods are not yet applied everywhere,
they are now well-known. Henceforth, we shall not develop the
subject any further.

The basic control method remains the ion-exchange. The
applications of the method in metal finishing shops are numerous
and they are now well commanded. Ion-exchanges are used for two
purposes:

. reclaiming of solutions by the fixation of the metallic
pollutants on resins, thus preventing discharge.
. reintroduction of drag-outs, by fixation and concentration on ion-exchangers.

Ion-exchangers can be used whatever the size of the shops
and the effluent flows may be. However, in spite of recent technological progress, some applications still need improvements such
as concentration of very acid solutions when a capacity to concentrate effluents remains limited. As regards these applications,
ion-exchangers are now being replaced with new technologies among
which:

. evaporation and electrodialysis applied to effluent concentration,
. electrolysis and electro-electrodialysis applied to effluent
  purification.

Evaporation which allows the dewatering of diluted effluents
before their re-introduction into the plating bath, is highly
recommended, when part of the energy required is provided by the
condensation of the plating bath vapours. This process is not employed
much yet, possibly because of its high investment and exploitation
costs, and its being chemical engineering which limits its implementation to large-scale shops.

Electrolysis is a well-known widespread method, combining
low operation cost with satisfactory energetic performances. It can
be used in order to reclaim in a metallic state the additive salts
contained in waste baths or in waste rinse waters. In the case of
closed circuit, this allows maintenance of a constant content in
impurities. The recovered metals will be re-introduced into the
plating bath, as anodes. New electrolysis equipment satisfactorily
solves difficult problems such as the nickel electrolysis low
efficiency.

Electrodialysis which combines the entrainment effect of
the ions stimulated by electric current with the piling of
alternate selectivity ion-exchanging membranes is of great
utility in the concentration of waste rinse waters to be later
reintroduced into the plating baths.

Electrodialysis is very efficient in the concentration
series which can hardly be tolerated by ion-exchangers. This
method which has proved to be very satisfactory in other fields
such as sea-water desalinization and lactoserum demineralization,
has already been tested in a few metal-finishing shops, and will
expand rapidly.

Electro-electrodialysis associates on the electrodes electrolysis and electrodialysis reactions. It is applied, for example,
to low-grade effluent demetallization. In these concentration
series, electro-electrodialysis is more efficient than ion-
exchangers. It is also applied to the purification and dewatering
of chromium effluents for which electro-electrodialysis is more
performing than ion-exhangers. These methods, elaborated in order

to solve control problems in metal-finishing shops, are well adapted to these problems and will develop rapidly.

Other membrane methods, essentially reverse osmosis, particularly for concentration of rinse water containing nickel and ultra-filtration, could also be extended.

It should also be mentioned that research goes on and that new bath technology offers a variety of methods which allow, if compatible with quality requirements, for a decrease in the volume, the concentration, and/or the toxicity of effluents. Technological substitutions can go as far as a complete change in the nature of the applied treatment, e.g., the vacuum metallization process.

Recovery which is closely related to control allows for a valorization of the metals contained in the concentrated effluents, which are mainly represented by waste plating baths. These metallic salts proceed either from additives or from the chemical attack of the substrate.

Technological progress provides industrialists with methods allowing for the recovery of most of the metals contained in effluents, but the main barrier to their valorization is the need for commercial outlets.

The treatment of effluents

Whatever may be the results of the additive loss control policy, standard metal finishing shops will still produce effluents that will have to be eliminated, and which are too often landfilled when the regulations in force require a pre-treatment process.

Diluted effluents can be treated either in "open circuit", i.e., with concentrated effluents or in "closed circuits". In the latter case, ion-exchangers allow concentration of the effluents to obtain pure, often demineralized, recyclable water. This treatment is usually chosen for the more diluted effluents or when their treatment is achieved in a specialized collective center with mobile exchangers.

When necessary, cyanide should be removed from effluents, (oxidation of cyanides to cyanates, and rarely, degradation in nitrogen and carbon dioxide) and chromium should be removed as well, (reduction of hexavalent chromium to trivalent chromium). Effluents are then neutralized and metallic salts are precipitated as non-soluble hydroxides. Sludges are then separated by clarification dried to a variable water content, and landfilled.

However this treatment is a conditioning rather than a destruction of hazardous wastes. To resort to a collective treatment center is possible at different stages of the neutralization
process. It seems, for example, that some shops that may be equipped
for cyanide removal and/or for chromium removal are not equipped
for non-soluble hydroxide precipitation.

Experience proves that the annual operation costs af an
effluent treatment station usually range from 10 to 20 % of the
total investment (tax and financial cost free). A saving policy
on investment will have repercussions on operation costs. Moreover
neutralization collective facilities foster the implementation
of programs of the following type:"numerous rinses + concentration".
To resort to these collective facilities entails a transfer of
investments from the following parts-- cyanide removal, chromium
removal, pH adjustment, flocculation, clarification, sludge
dewatering-- to the concentrating parts. Since dewatering is by
far the most important part among the following remaining ones
(storage areas, collecting pipe-system, final neutralization), this
transfer emphasizes the need for a reduction of the volume of the
effluents to be treated which will be obtained by control efficiency.

RECOMMENDATIONS

The effort to reduce pollution from metal finishing industry takes place at two levels :

    . in priority, development of pollution control techniques,
    . satisfactory treatment of wastes inevitably produced.

Development of pollution control techniques

These techniques exist, but must be perfected and adapted
to each case. In order to develop their use, it is necessary
to change the methods and outlook of the manufacturers in question.
Several types of measures may be recommended to this end.

    Regulation pressure

Waste laws and executive enactments prescribe on metal
finishing manufacturers many obligations which they feel are
constraints. But regulation may be used as a pedagogic tool
to urge manufacturers to employ less polluting processes and
to increase their recycling and recovery potential. For example,
these rules could encourage the establishment of registers in
which all consumptions, flows, fluxes, discharges would be systematically noted.

Moreover, regulations could allow for a clear identification
of responsibilities within companies involved. The precise designation of a "Waste Manager" within the company, with a sufficient
level of responsibility, is a basic element for a detailed study
of means to be used in the fight against pollution either at the
waste pollution level or at the treatment level. (Clause 20 of the
Belgium Waste Act of the 22nd July of 1974 requires designation of
a person responsible for waste treatment, as does German legislation).
The Waste Manager must communicate directly with the authorities.
Manufacturers would bear the cost of renumeration and training of
the waste manager.

Training and Information

The various "Waste managers" must be trained in new techniques. It is thus necessary :

. to disseminate detailed technical information documents,
. to develop specialized professional training in order to
  inform shop managers of practical aspects of handling of
  raw materials (setting up an input-output balance and
  an economic balance),
. to provide for standard guidelines.

Economic pressure

It is desirable to apply the "polluter pays" principle,
whereby pollution producers pay taxes or fees as a function of
quantities consumed or discharged. In order that measure be
applicable, discharges standards must be related to a quantitative
control system.

Applied research and consulting

Research for new simple technologies which can be easily
used must be continued. The adaptation of these new technologies
to various particular cases must be fostered. In this context,
it appears worthwhile to encourage the creation of a "technical
centre" (or the creation of a division in an existing research
organization which will be charged with these problems). The
purpose of this organization would be, in addition to the development of loss control techniques, to technically aid industries
and to organize the necessary training.

Recovery

Recovery which allows for marketing the main substances
contained in operation residues is an important means to be
used in fighting against pollution (and in saving raw materials).
It is thus necessary to aid the creation of facilities specia-

lizedin recovery and to encourage those technologies that have
significant impact on the potential for recovery.

It will always be necessary to remind people that it is inad-
visable to mix various waste streams if recycling and recovery
possibilities are to be improved.

Effluent and sludge treatment

Regardless of the reason, wastes are produced which must
be satisfactorily treated by physico-chemical treatment after
recovery in concentrated baths when possible.

A certain number of recommendations may be made to assure
this treatment under the best possible conditions:

. It is recommended, especially for small capacity units,
to resort to a specialized collective treatment centre or,
where this is not possible, to use the facilities of greater
units. This would assure a greater economic efficiency and,
more importantly, greater safety by enabling use of recovery
solutions provided that the various effluents or sludges are
not mixed.

In the case of individual industrial waste treatment facilities,
appropriate personnel training is necessary to achieve a satisfactory
operation of the station.

. In the case of liquid waste discharge to public sewage
treatment plants, it should be ensured that the following is
respected :
  - no disturbance to the operation of the waste water treat-
    ment plant,
  - no contamination of the sludge issued from the waste water
    treatment plant,
  - no intolerable release into the environment.

. Temporary storage of sludges containing highly valuable
materials for which recovery technologies could be developed
during the next years, is a solution to encourage. However, this
storage will have to be selective according to what the sludges
contain, and will have to be carried out under satisfactory conditions,
with regard to the environment's protection.

ACKNOWLEDGMENTS

We thank the experts from Canada, the United States, the Netherlands, the Federal Republic of Germany and the United Kingdom who have answered various questionnaires which were addressed to them and who have provided us with an abundant technical documentation.

BIBLIOGRAPHY

NATO/CCMS - 1976 - Traitement des déchets dangereux
Le cycle du chrome N° 52

Ministère de l'Industrie et de la Recherche (France) - 1976
SERATRADI : Le cycle du chrome et les possibilités de son
recyclage

NATO/CCMS - 1981 - Disposal of hazardous wastes-
Metal finishing wastes N° 121

Ministère de la Qualité de la Vie (France) - 1976 - CEPLAM
Lutte contre les eaux usées et déchets spécifiques de traitements
de surface

Ministère de l'Environnement et du Cadre de Vie (France)-1979 -
FAIRTEC - Les techniques de prévention dans les ateliers de
traitements de surface

Environmental Protection Agency (United States) - 1978
Development document for proposed existing source pretreatment standards for the electroplating point source
category 440/I -78/085

Department of the Environment (United Kingdom) - 1976
Waste management Paper N° 11 - Metal Finishing Wastes :
a technical memorandum on arisings, treatment and disposal
including a code of practice

Fisheries and Environment Canada - Water pollution Control
Directorate (Canada) - 1977
Metal finishing liquid effluents guidelines EPS 1 - WP 77.5

Environment Canada - Water pollution control
Directorate (Canada) - 1976
Waste water and sludge control in the canadian metal finis-
hing industry.

HAZARDOUS WASTE REDUCTION AND RECYCLING

Dr. K.W. Riegel

U.S. Environmental Protection Agency
Washington, D.C., United States

"Pollution is nothing but the resources we are not harvesting.
We allow them to disperse because we've been ignorant of their
value."

—Buckminster Fuller

I. INTRODUCTION

Although almost every waste has the potential for causing
environmental damage, as a practical matter it has been estimated
that 10-15[1] percent of the industrial wastes generated in our
modern industrialized society may be classified as hazardous.
That is to say, these are wastes which pose special hazards to
public health and the environment unless they are properly
handled, treated, stored, transported and disposed of.

As a general rule, hazardous wastes are those which contain
toxic chemicals, explosive substances, flammables, acids, caus-
tics, pesticides, radioactive materials, infectious agents, or
other substances in sufficient amounts to be capable of causing

---

[1] National Solid Wastes Management Association, " Mismanagement is
the Real Hazard in Hazardous Wastes," April 1980.

acute or chronic health effects or severe environmental insults.
It is utopian to expect that even the most advanced technology,
enlightened management, and farsighted legislation can eliminate
all hazards from our waste streams. However, it is possible to
curtail health and environmental damage caused by hazardous wastes
by reducing such waste and recovering and recycling valuable
resources. It is equally possible to modify certain industrial
processes so that hazardous by-products are not produced or are
produced in diminished quantities. We also can treat some of
these wastes so that the hazard is neutralized.

Why, then, did the United States not address the hazardous
waste problem long ago? Environmental ignorance is part of the
answer. We didn't know that most hazardous wastes were hazardous
and we were not looking for the problem. Economics furnished the
rest of the answer. In the past, when energy, virgin materials,
and land were abundant and inexpensive, little thought was given
to recycle/reuse technologies. Government policy and industrial
practices generally favored environmentally unsound disposal of
wastes, hazardous or not, because the costs of environmental and
human health damage did not appear in the expense ledgers of business or government.

Such externalities have, unfortunately, begun appearing in
our ledgers, and the prospect of very large future costs of this
kind has motivated public and private action to improve on past
practice. Acceptable means for dealing with hazardous wastes have
been around for many years, but the internal cost of these alternatives often exceeded the cost of environmentally damaging
methods. In our highly competitive business atmosphere, companies
in an industry which sought to use environmentally acceptable dis-
posable methods found themselves at a serious competitive disadvantage.

By the early 1960's the stage was set for change. Rachel
Carson's book, Silent Spring, alerted scientists, government officials, and the general public to the fact that a substance introduced into one part of the environment could move by various and
complex pathways that were not well understood. We began to
understand that there could be serious consequences even from
well-intentioned uses of toxic materials. Throughout the 1960's
and into the 1970's concern mounted, and both government and
industry sought to achieve greater environmental responsibility.
Air and water pollution were quickly addressed as was misuse and
abuse of pesticides. As awareness expanded, toxic substances in
general became of concern as did the disposal of hazardous wastes.

The Clean Air Act, the Federal Water Pollution Control Act,
the Federal Insecticide, Fungicide and Rodenticide Act, the Toxic

Substances Control Act, the Resource Conservation and Recovery Act, and the Comprehensive Environmental Response, Compensation, and Liability Act (Superfund) were among the legislative responses by the U.S. Congress to public and scientific environmental and health concerns. These laws and their attendant regulations provided legal and economic incentives to re-evaluate the generation and disposition of hazardous materials. Environmental damage became a cost factor to be added to the price of doing business. To this have been added the rapidly escalating costs of virgin materials and energy, scarcity of many resources, and public hostility to locating hazardous waste disposal sites in nearly any community.

Among the clear responses to this issue is the possibility for reducing or recycling hazardous wastes. Wastes not generated, or wastes recycled by confinement or neutralization to harmless and useful substances, are removed from the inventory of hazardous material that must be dealt with. The attractivensss of this approach, where practical, is obvious. This paper will present some highlights of approaches in the United States that show promise.

II. SOME CASE STUDIES OF ACTUAL RECYCLING APPROACHES EMPLOYED IN THE UNITED STATES

3M

Minnesota Mining and Manufacturing Company (3M) is a multi-national corporation based in the United States. In 1976, 3M decided to offset the costs of pollution control by employing innovative technology to turn pollutants into resources and new profits. The program became known as "3P" - "Pollution Prevention Pays."[2] In its first nine months, 3P eliminated 500 million gallons of wastewater as well as 70,000 tons of air pollutants for a savings of $11 million. In its first three years, 3P has saved over $20 million in its operations. Solid waste, much of which could be considered toxic, has been cut back from 6,000 tons per year to 1,800 tons. 3M's organizational structure for combating waste is particularly interesting. Each 3M plant has an energy and environment committee made up of the plant engineer (chairman), the manufacturing supervisor, the process engineer, the control engineer, an industrial engineer, and the maintenance supervisor. This committee establishes goals for the reduction and elimination of waste, works out programs with shop personnel and reports progress to management. All plant personnel are

---

[2] Rayston, Michael G. "Making Pollution Prevention Pay," Harvard Business Review, November-December 1980, pp 6-14.

encouraged to furnish knowledge, know-how, and insights to this
program. The specific foci of 3P are to reduce or eliminate waste
and pollution through more efficient use of raw materials and
energy, to redesign or modify processes to reduce waste and pollution, and to find ways of recovering and reusing wastes and byproducts.

Uniroyal

Uniroyal Chemical has won national recognition for developing
an environmentally acceptable way of disposing of nonenes, an
extremely volatile, hazardous waste left over from the manufacture
of anti-oxidant rubber chemicals. Instead of shipping nonenes to a
disposal site, Uniroyal mixes them 50-50 with fuel oil and burns
the mixture in the company's steam-generating plant. This has
resulted in lower fuel costs, elimination of shipment expenses and
reduced employee exposure to a volatile liquid. Equipment costs
amounted to $48,000. However, the first year savings alone
amounted to $183,000.[3]

Foremost-McKesson

Foremost-McKesson, a San Francisco based firm, is engaged in
a thorough analysis of the chemical recycling industry which pro-
mises to reclaim approximately $100 million in new revenue from
recycled chlorinated and fluorocarbon solvents. The first of
possibly seven Foremost-McKesson Chemical Group plants is an
expanded and redesigned version of a pilot facility located in the
Chicago area. During 1979, 1-1/2 million gallons of solvents were
recycled. The expanded plant is expected to increase that capacity by 30 percent.[4] Approximately half of this company's chemical business is devoted to sale and distribution of virgin solvents for cleaning electronic circuitry and metal parts, and
hydrocarbon solvents for making paint and ink. The new approach
calls for the company to collect the used solvents, remove the
impurities and resell the recycled solvents. In all likelihood
these recycled solvents will be sold back to the companies who
originally purchased them, but at a significant cost savings to
both seller and buyer.

OIL SALVAGING

Petroleum manufacturers and processors have for some time re-
refined and reprocessed dirty and used oils. Reprocessing is the
less complicated of the two operations. The source of waste oil

[3] ChemEcology, News Item, April 1979, p. 7.

[4] Chemical Purchasing, News Article, May, 1981, p. 85.

collected is automobiles, small power boats, agricultural and
industrial lubrication, industrial "slop" oil from fuel storage
tank cleaning, and the like. In reprocessing, the collected waste
oils are dewatered, sediments and certain other impurities are
removed, and the reprocessed oil is mixed with virgin oil to produce a fuel oil rather than a lubricating oil. It is a notably
less expensive fuel than virgin oil alone.

A commonly employed process for re-refining waste oil is the
acid/clay treatment process in which waste oil is mixed with sul-
furic acid and clay to remove impurities. The oil/clay mix is
heated, then cooled; the clay containing impurities is then
removed by filtration. The remaining oil is mixed with additives
or with virgin oil and may be reused as an industrial lubricating
oil. Unfortunately, the clay sludge must be disposed of in a
secure landfill. In this process, the hazard is reduced in volume
but not eliminated.

Another re-refining process which shows considerable promise
is the distillation/hydrotreating process. As in the acid/clay
process, debris is filtered out of used oil, water is evaporated
by heat treatment, and light oil is separated and recovered for
use as a fuel. Here the similiarity ends. In the distilla-
tion/hydrotreating process, the remaining waste oil is subjected
to certrifuging which removes heavy impurities. The waste oil is
then piped to a vacuum tower where lubricating oil is vaporized,
then recovered for industrial uses. The remaining impurities are
mixed with the centrifuge residues and the mixture is used or sold
as a low-grade fuel. In this process, the hazardous clay/impur-
ities sludge is eliminated.

CHEM-METALS

Recently Chem-Metals, Inc., an Indiana based firm, announced
plans for recovering and reusing nearly 100 percent of hazardous
solder dross skimmings from its can manufacturing operations.
This company also intends to employ their patented hydro-metallurgical system on tin, lead, and zinc drosses produced by electroplating, lead smelting, rubber and automobile-making. A company spokesman explained that the hydro-metallurgical process
recycles solder dross skimmings into solder ingots and transforms
spent dross into chemicals that are compounded to produce
materials like roll salts.[5]

---

[5] Solid Waste Report, News Item, July 1, 1981, p. 101.

LAND TREATMENT

Petroleum producers such as Exxon and Continental Oil have
been disposing of sludges by land treatment techniques for more
than 10 years. Biodegradable sludge is spread on land to permit
water to evaporate, and is then disked into the soil. After about
two months, the oil content of soil has been reduced by 96 to 98
percent and another application can be made. Of course, such
sludges must have a negligible heavy metal content and the topographic and climatic conditions must inhibit erosion and runoff.[6]

CORPORATE WASTE MANAGEMENT STRATEGY CHANGES

Many other companies such as Hercules Powder, Hanes Dye and
Finishing Company, Dow Corning, and Westvaco have demonstrated the
profitability of resource recovery from waste streams. From these
examples, we may conclude that in many industries much waste and
pollution derives from process inefficiency. We may further conclude that by applying the philosophy that pollutants can become
potential resources, with know-how and imagination.

These examples are some of the company initiatives recently
taken to reduce hazardous wastes and reclaim valuable resources.
There are doubtless many further opportunities, although, in some
cases individual companies may be reluctant to disclose their proprietary recycling/reuse processes.

III. GENERIC APPROACHES TO WASTE MANAGEMENT

Intra-company reduction and reuse of wastes is vital, but
there are many industrial byproducts which a given generator cannot reuse, recycle, or burn for energy. However, one company's
waste may be another company's resource. This observation has led
to the development of waste exchanges. Formal waste exchange
systems first blossomed in Europe and evolved later in the U.S. in
several ways. The European approach has tended to center on
information exchanges, whereas some U.S. exchanges have gone so
far as to function as brokers, or even to obtain materials for
reprocessing and resale. At least two American exchanges provide
exchange services as part of overall waste management consultant
services.

---

[6] Maugh, Thomas, H. "Hazardous Wastes Technology Is Available,"
Science Vol. 204, June, 1979.

DEVELOPING WASTE MANAGEMENT STRATEGIES

As a first step in institutionalizing the concept of waste exchanges, the Environmental Protection Agency, in 1976, suggested to U.S. industries that plant managers and engineers consider the following sequence of steps as they develop their waste management strategies:

1. Minimize the quantity of waste generated by modifying the industrial process involved.
2. Concentrate the waste at the source to reduce handling and transport costs.
3. If possible, transfer the waste "as is" without processing, to another facility that can use it as a feedstock.
4. When a transfer "as is" is not possible, reprocess the waste for material recovery.
5. When material recovery is not possible: (a) incinerate the waste for energy recovery and for destruction of hazardous components, or (b) if the waste cannot be incinerated, detoxify and neutralize it through chemical treatment.
6. Use carefully controlled land disposal only for that which remains.

IN PRACTICE, HOW HAVE THINGS WORKED OUT?

In the information exchange, the clearinghouse has become the cornerstone of the system, with the generator and potential user forming the other two points of a triangular relationship. The entire system is built on a foundation of confidentiality. Confidentiality is best maintained when the operation is controlled by a non-regulatory body (Chamber of Commerce, trade association, etc.). Industry is reluctant to reveal detailed information about a waste stream for two reasons:

1. to avoid alerting competitors to proprietary information which gives a generator an economic advantage, or
2. to avoid a regulatory body using "inside information" against them, regardless of all good intentions.

STEPS IN A SUCCESSFUL EXCHANGE

Once a generator has identified a waste stream with potential value, he will list it with the clearinghouse.

The clearinghouse then codes the identity of the generator and lists the pertinent information. The clearinghouse follows a similar procedure for companies that request specific waste materials. In each case, confidentiality is maintained. Once the

listings have been compiled, the clearinghouse will issue a peri-
odic publication to match up "donors" and "consumers."

Upon finding a listing of a material which may be useful, the
potential user will inquire as to the availability of this material. Quite naturally, some of the most "appealing" wastes are
those which are available on a regular basis and in sufficient
volume to justify use.

Now two sides of the triangle are complete; the generator has
listed with the clearinghouse and the potential user has inquired.
To complete the triangle, the clearinghouse notifies the generator
of the interest expressed in his material. The clearinghouse then
steps aside and leaves final negotiations to the generator and the
potential user.

MATERIAL EXCHANGE

In this system, the material exchange takes physical posses-
sion of the material. It may analyze, treat, and/or store the
material before reselling it at a profit. By nature, the materials exchange requires a greater capital investment, more extensive technical expertise, and more aggressive marketing.

EXCHANGES IN OPERATION

The first information exchange was established in St. Louis,
Missouri, in 1976. This was followed by clearinghouse exchanges in
Indiana, New Jersey, North Carolina, Virginia, Oklahoma, New York,
Arkansas, Pennsylvania, and Michigan. Boston, Massachusetts, was
the home of one of the first materials exchanges with others
following in California, Illinois, Michigan, Ohio, and New York, to
name a few. There are two computerized exchanges in the U.S.; one
in Minnesota and one in Connecticut. Clearly, computer handling
of large volumes of waste exchange information has national and
international applications in the future.

WASTE EXCHANGE

Not all exchanges have become instant successes. Some have failed, some have had to reorganize, and others have grown slowly.

LESSONS LEARNED

Wastes most likely to be exchanged are acids, catalysts, solvents, combustibles, residues with high metal content, and oil. The most likely exchange partners have been large companies with small companies, chemical manufacturers with formulators, and high purity industry (such as pharmaceuticals) with low purity industry (such as paints).

High transportation costs tend to inhibit trades beyond fifty miles as a general rule. This varies according to local disposal fees, the intrinsic value of the waste, and the impact of local and Federal laws. Low concentration of useful materials and impurities in the waste can also be deterrents. These may be alleviated by waste segregation, altering processes or materials changes at the original manufacturing site.

IV. SOME EPA TECHNICAL INITIATIVES TO REDUCE WASTE

I will now discuss a few examples of EPA's efforts to develop information, processes and pilot projects aimed at helping industries and exchanges involved in hazardous waste reduction, reuse, and recycling.

The electroplating industry produces hazardous wastes made up of heavy metals and other chemicals. These wastes can become among the more serious environmental and health hazards, and yet they contain many valuable byproducts. This industry has an estimated 20,000 shops, the vast majority of which have insufficient capital and expertise to purchase and effectively operate wastewater pollution control equipment.

EPA, through its industrial research laboratories and grant programs, is examining cost efficient technologies for extracting metals, plating solutions, and various toxic chemicals from industrial waste streams.

HSA REACTOR

Present treatment of electroplating wastewaters involves a two-stage process for cyanide destruction, the precipitation of heavy metals, and subsequent land-disposal of sludge. These practices not only generate large wastewater flows (1 billion gallons per day - 4.545 billion liters per day - industry wide), but also reject metals which could be reused.

EPA is about to complete a demonstration study of a full--
scale electrochemical HSA (high surface area) reactor under a cost
sharing agreement with a metal finishing company in Cincinnati,
Ohio. The newly developed reactor destroys cyanide and deposits
heavy metals on large-surface carbon fiber cathodes. The cathodes
are then periodically stripped for metal recovery and reuse in the
plating baths. This treatment technology also allows the recycling of wastewater in the rinse cycles. A process blowdown is provided to control dissolved solids. Other potential benefits arising from the use of this system include reduced energy use and
easier National Pollutant Discharge Elimination System (NPDES)
compliance, reduction in the generation of metal-bearing sludge,
low installation costs, reasonable capital costs, and relative
simplicity of operation. Cost figures and operating data from our
tests will soon be available.

ION EXCHANGE

Ion exchange is a versatile separation process with the
potential for broad application in the metal finishing industry,
both for raw material recovery and for water pollution control.
Ion exchange technology has been greatly enhanced by the broad
range of resins currently being manufactured. With proper resin
selection, ion exchange can provide an efficient and cost-effective solution to waste reduction requirements.

Our investigations have shown that there are three areas of
ion exchange application. In the first application, mixed rinse
solutions are de-ionized to permit reuse of the treated water.
The contaminants in the rinses are concentrated and are made more
economical to treat. This application results in a significant
reduction in the volume of wastewater discharged and in total
water consumed, thus reducing water use, sewer fees, and the size
and cost of the pollution control system.

In the second application, toxic heavy metals and metal cyanide complexes are removed selectively from combined waste streams
prior to discharge. The key to this application is that the ion
exchange resins remove only the toxic compounds while leaving non-
toxic dissolved ionic solids to remain in solution.

In the chemical recovery application, segregated plating
rinse waters are treated to concentrate the plating chemicals for
recycling into the plating bath. The purified rinse water is also
recycled. In general, ion exchange systems have been found suitable for chemical recovery applications where the rinse water feed
has a relatively dilute concentration of plating chemicals and a
relatively low degree of concentration is required for recycle of

the concentrate. Ion exchange is well suited for processing corrosive solutions. Ion exchange has been demonstrated commercially
for recovery of plating chemicals for acid-copper, acid-zinc,
nickel, tin, cobalt, and chromium plating baths. It has also been
used to recover spent acid solutions and to purify plating solutions for longer service life.[7]

CASE STUDY

The Hurd Lock and Manufacturing Company of Greenville,
Tennessee, employs a combination treatment system using batch
treatment of wastewater collected in four different treatment
sumps. The wastewater collected in each sump is processed through
the system as shown in the following schematic.

The continuous treatment system for each batch employs the following steps:

o   Chromium reduction (not always required)

o   Neutralization (the pH set-point adjusted to achieve max-
     imum metal removal for each waste processed)

o   Flocculation (with polymer addition)

o   Pressure filtration (with diatomaceous earth precoat)

o   Ion exchange polishing

---

[7] U.S. EPA, Environmental Research Laboratory, Cincinnati, Summary
    Report, "Control and Treatment Technology for the Metal
    Finishing."

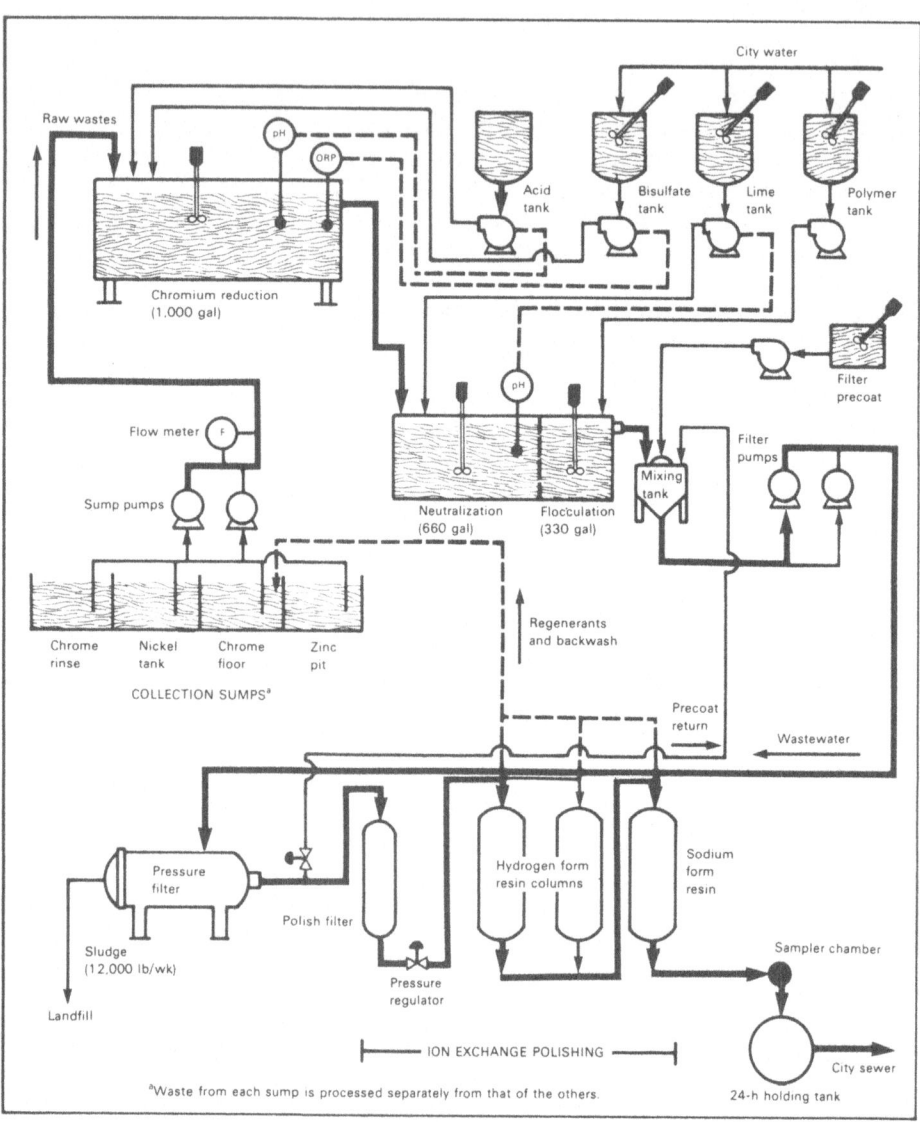

Continuous Treatment System

The ion exchange polishing step reduces the concentration of the metals to the level required for discharge. The following table presents the discharge quality and the requirements set forth in the permit:

Permit Requirements and Treated Effluent Quality: Ion Exchange

Effluent characteristic	Permit requirements	Treated Effluent			
		Chrome floor	Chrome rinses	Nickel rinses	Zinc pit
pH	6.5–8.5	11.0	11.0	6.9	11.6
Color units	12	0	0	0	0
Pollutant (mg/l):					
Total suspended solids	15	*1.0	*1.0	*1.0	*1.0
COD	20	928	210	217	500
Cadmium	0.01	*0.005	*0.005	*0.005	*0.005
Chromium	0.05	*0.02	*0.02	*0.02	*0.02
Copper	0.05	*0.05	*0.05	*0.05	*0.05
Iron	0.50	*0.05	*0.05	*0.05	*0.05
Lead	0.05	*0.05	*0.05	*0.05	*0.05
Nickel	0.10	*0.05	*0.05	*0.05	*0.05
Zinc	0.10	*0.02	*0.02	*0.02	*0.02

*=Less Than

Note: The plant was allowed to exceed the chemical oxygen demand (COD) level because the pollutant could be effectively reduced by subsequent treatment.

MEMBRANE SEPARATION OF METALS

Another promising metal recovery process is the use of coupled transport membranes. It has been experimentally demonstrated that metal ions can be "chemically pumped" across a coupled transport membrane against large concentration gradients by allowing the counterflow of a coupled ion such as hydrogen ion. The process is carried out within a microporous membrane containing within its pores an organic, water-immiscible complexing agent. The complexing agent acts as a shuttle, picking up metal ions on one side of the membrane, carrying them across the membrane as a complex, and preserving electrical neutrality by carrying hydrogen ions in the opposite direction. At this stage of development, the economics of such a recovery system appear favorable.[8]

[8] EPA, Coupled Transport Systems for Control of Heavy Metal Pollutants, August, 1979.

CASE STUDY

Under an EPA-sponsored demonstration program conducted by Bend Research, Inc., Bend, Oregon, the following results were obtained:

Chromic Acid Recovered   –   4900   micrograms/cm^2

@ 60¢ lb. (.0013¢/gram)   – $26.00 per ft^2 ($280/m^2) of membrane/year

Membrane cost   –   Approx. $10/ft^2 ($108/m^2)

Thus, the expected payback time for the membrane concentrator, ignoring operating costs, could be less than 6 months.

Under favorable circumstances where complete recyling is practiced, there would be an additional savings in the cost of make-up water for the rinse bath. This could amount to a savings of 1000 gallons (3785 liters) of water per ft^2 (0.093 m) of membrane per year, or approximately $1/ft^2 ($11/m^2) of membrane per year.

In spite of the lower fluxes observed for copper and the lower cost of this metal, the economic outlook for copper is also quite favorable. Only for nickel is the current economic projection unfavorable because of low fluxes observed to date.

CENTRALIZED TREATMENT

EPA has also conducted industrial feasibility studies to aid the electroplating industry identify methods of complying with wastewater pretreatment and resource conservation and recovery regulations. We found an economically and technically attractive alternative to individual plant, on-site treatment, in the centralized treatment facility concept. Centralized treatment results in considerable savings, which are reflected in reduced costs of compliance for individual platers. Our findings, based in part on the Federal Republic of Germany's experiences, show that centralized treatment, in one or more of its forms, can provide a feasible alternative to expensive on-site treatment for a wide variety of plating shops. The central treatment facility should also make resource recovery more economic and/or lend itself to the waste exchange concept more effectively than the single plant operation.[9]

LEAD EMISSION REDUCTION

Lead, one of the most useful metals in modern industrial
society, has long been known to be highly toxic. Because of this,
emissions at smelter plants must be tightly regulated. In order
to discover cost-effective means of reducing lead emissions at
smelters, an interagency study of the Danish technology was undertaken. The National Institute of Occupational Safety and Health
and EPA found that the processes at the Paul Bergsoe and Son
smelter in Glostrup, Denmark offered potential solutions to major
U.S. occupational and environmental problems associated with
secondary crude lead production.

The EPA Industrial Research Laboratory, Cincinnati, then initiated a demonstration project with a secondary lead smelter using
the Bergsoe Technology coupled with domestically developed technological innovations. The demonstration showed that occupational
exposure to lead and atmospheric emissions could be substantially
reduced by installing these combined control technologies. The
data gathered will aid local enforcement agencies in recommending
low-cost alternatives to the capital intensive traditional
approaches.

TEST RESULTS

Smelter Feed Materials:

12.6%	polypropylene (case - whole battery)
12.6%	hard rubber (case - whole battery)
31.5%	battery plates
22.1%	return slag
6.3%	mill scale (FeO)
5.7%	coke
3.2%	agglomerated dust (or battery mud)
3.2%	drosses
1.9%	scrap iron
.9%	$CaCO_3$

Total 100.0%

---

[9] U.S. EPA, Industrial Research Laboratory, Cincinnati, Environ-
mental Pollution Control Alternatives - Centralized Waste
Treatment Alternatives for the Electroplating Industry,
June 1981.

Smelter Output Streams

	Percent of total incoming flow				
	Pb	Sb	As	Cl	S
Lead bullion	87.7	98.5	10.4	12.8	0.14
Lead stone (matte)	0.90	11.2	75.7	36.8	98.0
Stack gas	0.0025	0.84	0.07	75.8	7.2
Total	89	110	86	125	105

(Estimated accuracy of elemental flow rates is $\pm$ 20%.)

Average Production Rate

70.5 metric tons of lead per day

Controlled Stack Emissions Achieved

	Test Day 1		Test Day 2	
Element	Concentration (ug/Nm^3)	Emission rate (kg/hr)	Concentration (ug/Nm^3)	Emission rate (kg/hr)
Lead	1010	0.12	350	0.04
Antimony	4390	0.54	4370	0.52
Arsenic	4	0.0005	11	0.0013
Chlorine	13300	1.6	59500	7.1
Sulfur	55700	6.7	77100	9.1

Employee exposures were maintained below 100 ug/m^3 in all areas of the smelter. This low exposure is due to the exemplary engineering and work practice controls. Yard sprinkling, washdown procedures, and good general housekeeping efforts also helped reduce the levels of lead in and around the workplace.

CENTERS OF EXCELLENCE

Beyond the immediate technical needs of industry and EPA is the necessity to advance the science of waste reduction and reuse. One of the objectives of the EPA's Office of Research and Development grant program is to stimulate scientific and technical research fundamental to pollution control advances. Among our programs is the recent establishment of a number of innovative research centers at competitively selected universities. Each center focuses on specific long-term environmental problems. The center concept is intended to establish a cadre of outstanding researchers who are capable of contributing to EPA's long-range research objectives and to provide the link between basic and applied research. These centers and their programs have a multi-media and multidisciplinary orientation. This autumn, we will establish a hazardous waste research center. In its initial years of operation, the center's focus will be on technology issues related to the disposal and clean-up of hazardous wastes. The center will encourage municipal and industrial cooperation and collaboration. Through such efforts, we hope to preclude future episodes such as Love Canal, Valley of the Drums, and kepone con-tamination. In a more positive sense, we hope to unlock many more mineral, chemical, and energy resources now bound up in our mount-ing waste flows.

V. CONCLUSION

The preceding experiences, efforts, and studies all indicate that the reduction and reuse of hazardous wastes are as complex as the waste streams themselves. But progress in this technically complicated field is being made. This symposium is a particularly valuable mechanism for sharing ideas, approaches, and experiences which will benefit all nations represented here.

As stewards of our resources and protectors of the environ-ment and public health, our nations are driven to manage wisely our resources. The old saying, "waste not, want not," remains environmentally and economically valid today. It is a strong admonition to examine and reexamine everything we call waste -- especially hazardous waste, and find better ways to reduce what we must discard, to dispose of hazardous materials with great envi-ronmental caution, and to reuse and recycle whenever and wherever possible.

I am looking forward to hearing the experiences of the other symposium participants.

SELECTED BIBLIOGRAPHY

American Petroleum Institute and National Petroleum Refiners Association. Solid Waste Practices Under RCRA In the Hydrocarbon Processing Industry: API-NPRA Conference, January 31-February 1, 1980.

U. S. Environmental Protection Agency. Electrolytic Treatment of Wastewater From Manufacturing and Machining Plants. EPA-600/2-80-134, June 1980.

-----, Environmental Pollution Control Alternatives: Centralized Waste Treatment Alternatives for the Electroplating Industry. EPA-625/5-81-07, June, 1981.

-----, Control and Treatment Technology for the Metal Finishing Industry -- Ion Exchange. EPA-625/8-81-007, June 1981.

-----, "Controlling Hazardous Wastes." EPA-600/8-80-017, May, 1980.

-----, Coupled Transport Systems for Control of Heavy Metal Pollutants. EPA-600/2-79-181, August, 1979.

-----, Environmental Regulations and Technology: The Electroplating Industry. EPA-625/10-80-001, August, 1980.

-----, Evaluation of Paul Bergsoe & Son Secondary Lead Smelter. EPA-600/2-80-022, January, 1980.

-----, Reverse Osmosis Field Test: Treatment of Watts Nickel Rinse Waters. EPA-600/2-77-039, February, 1977.

-----, Waste Exchanges - Background Information. SW-887.1, December 1980.

THE ACTIVITIES OF THE EUROPEAN COMMUNITY ON HAZARDOUS WASTE

Benno W.K. Risch

Commission of the European Community
Brussels, Belgium

There is no doubt that the toxic and dangerous waste arising out
of industrial activities is one of the major environmental problems and takes its place among the priority tasks of environmental policy.

Treatment and disposal of this toxic and dangerous waste is the
number one qualitative problem of waste management, not because
of the quantities involved but because of the particular hazards
attached to it - such as toxicity, health hazards, danger of
water pollution, the risk of infection, the danger of explosion,
of fire, of corrosion and the like. These hazards derive from
the particular properties of these substances which, even in tiny
quantities, are highly dangerous.

Some 20 million tonnes of such toxic and dangerous waste arise
every year in the European Community. It accounts for some 15 to
20% of the Community's annual arisings of industrial waste - an
estimated 120 million tonnes at present. Most toxic and dangerous
waste occurs in the chemical industry as unavoidable by-products
of industrial processes.

In the last few years there has been an above-average increase in
toxic and dangerous waste as regards both quantity and their
complexity. This is, in part, a direct consequence of a heighte-
ned concern with the environment and of measures prohibiting
uncontrolled disposal, discharge or tipping, as well as of
advances in chemical and physical knowledge and of identification
of the particular properties of this waste and the threat to the
environment it represents.

This has led to the "discovery" of more and more new forms of
dangerous waste.

A further disproportionate increase in the amount of dangerous
waste to be disposed of can be expected in the next few years,
since considerable additional amounts will arise through flue
gas purification at waste incineration plants, through sewage
treatment and through the fact that the dumping of waste at sea
will be rendered more difficult by international conventions on
keeping the seas clean.

Only in the medium and long term can we expect a gradual decline
in the amount of toxic and dangerous waste as industry adapts
itself to legal, administrative and technical requirements, as
the cost of treating and disposing of such waste rises and as
the extensive research and development endeavours in this field
begin to produce results.

Industry has already begun to react to these challenges by :

- replacing materials, processes and technologies, so as to
  reduce the amount of dangerous waste at the potential point
  of arising;

- introducing or applying low-waste technologies; this has
  largely eliminated hazards due to hardening salts;

- increasingly recycling dangerous waste for energy purposes or
  as secondary raw materials - this has been tried successfully,
  in particular with used solvents and waste from galvanizing
  processes.

However, the steady increase in the quantity of toxic and
dangerous waste means that an optimum disposal technology is
needed, which - on land in particular - is able to deal with
such waste in an economically and ecologically acceptable manner.
Particular efforts will have to be concentrated on improving
technical and organizational disposal and treatment methods,
with recycling technologies in support of the reduction and
disposal of waste having increasingly to be embodied in an
overall plan or strategy of waste management for hazardous waste.
For dangerous, as for other waste, there will have to be a switch
from disposal to recycling. There are three methods of waste
disposal and treatment: chemico-physical treatment, thermal
treatment and dumping. Optimized collection and transport systems, control and monitoring regulations and the improvement of
methods of chemical analysis are important supplements to these
methods.

Aware of the particular hazards and significance of the production and disposal of toxic and dangerous waste for the environ-

ment and for economic development, and bearing in mind the growing extent of transfrontier haulage of such waste between the Member States, the Action Programme of the European Community for Protection of the Environment assigned a particular priority to this form of waste in Community endeavours in the field of waste management.

In implementation of the Action Programme and the framework Directive on waste of 15 July 1975, the Commission drafted a separate Directive on toxic and dangerous waste, which was adopted by the Council on 20 March 1978. This Directive lays down that its provisions must be incorporated into national law by 20 March 1980 by the individual Member States; so it has been in force since that date.

Article 21 of the Directive requires the Member States to provide the Commission by that date with detailed information on how the individual provisions of the Directive were implemented. However, the Commission does not yet have a conclusive review of the extent to which the Directive has been implemented by the individual Member States, as they have not yet fully complied with their obligations under Article 21.

However, the information available to us indicates that in at least six of the nine Member States major provisions of the Directive are already in force.

The Directive on toxic and dangerous waste is of particular importance, since it lays down common rules and provisions in respect of the major problems of the production and disposal of toxic and dangerous waste in this important and priority area of waste management.

The Directive contains the first definition at Community level of toxic and dangerous waste in the form of an Annex listing the 27 most important groups of toxic and dangerous substance to which the Directive is applicable.

The definition provided by the Directive - which was the subject of a very long tussle with government and industrial experts from the Member States - is by no means scientifically incontestable. If it had been our ambition to arrive at a scientifically incontestable definition, there would have been wrestling with this problem for another ten years or even longer. In fact, we would have had to define the whole problem of toxicity.

The Commission therefore attempted to find a pragmatic solution in the form of a list of substances, so as to solve the practical problems with which all the Member States of the Community are faced.

The chief provisions of the Directive are :

- the prohibition of uncontrolled discharge, uncontrolled transport and uncontrolled treatment and dumping of toxic and dangerous waste;

- the establishment of appropriate labelling indicating the type, composition and quantity of waste;

- the identification of sites at which toxic and dangerous waste is or has been dumped and identification of such waste;

- the requirement of a licence for plants, installations and undertakings which store, treat and/or dump toxic and dangerous waste;

- the requirement that the owners of toxic and dangerous waste who are not authorized to treat or dump such waste must hand over the waste to authorized plants, installations or under-takings for harmless disposal;

- the requirement that the relevant authorities must draft and develop plans for the disposal of toxic and dangerous waste; these plans must provide for the necessary special treatment plants and suitable dumping sites; they must also be published;

- the requirement that all plants, installations or undertakings which produce, own and/or dispose of toxic and dangerous waste must keep a special record of the quantity, type, physical and chemical characteristics, origin, method of disposal, dumping site and arrival and departure dates of such waste;

- the requirement that where toxic and dangerous waste is trans-ported in the course of disposal it must be accompanied by a special identification form until its final harmless disposal; these forms must be preserved;

- the requirement that every three years the Member States must draw up a report on the disposal of toxic and dangerous waste in their respective countries and forward it to the Commission.

The Commission circulates this report among the other Member States and reports to the Council and the European Parliament every three years on the application of the Directive. The first report has to be submitted in 1981. We hope that the Member States will inform the Commission comprehensively and in good time, so that the Commission will be able to comply with its obligations to the Council and Parliament.

The Directive of 20 March 1978 on toxic and dangerous waste in the Community was initially designed as a specific implementing Directive for the framework Directive of 15 July 1975. However, in the course of consultations with government experts from the Member States within Council committees it became, in its turn, only a kind of framework Directive for the toxic and dangerous waste sector. A series of highly specific rules in the original Commission proposal fell by the wayside in the course of the consultations with the government experts. This - along with the complex nature of this waste - means that, as Phase 2, the framework Directive on toxic and dangerous waste will have to be supplemented and amplified.

For Phase 2 of the Community's endeavours in the toxic and dangerous waste sector the Commission, along with a working party of government and industrial experts, has drawn up a medium-term programme.

This programme is composed of two parts: "priority tasks and measures" and "other problems".

The following points are listed among the priority tasks and measures:

1. Examination of identification and accompanying forms relating to the transport of toxic and dangerous waste pursuant to Article 14(2) of the Directive, particularly in respect of transfrontier carriage.

2. Examination of documents relating to special regulations, particularly in respect of labelling, and to safety precautions in respect of hazards, accidents, etc.

3. Examination of criteria for the monitoring and management of the final dumping or longer-term dumping of toxic and dangerous waste. The drawing up of special storage plans.

4. Transport problems.

5. Examination of the question whether - and if so which - concentrations should be determined for the substances listed in the Annex to the Directive with a view to possible proposals to the Council.

6. Particular problems of recycling, treating and disposing of certain forms of waste, such as :

   - organic halogen compounds;
   - used solvents;
   - waste derived from surface treatment;
   - arsenic;

   for which special codes of practice are to be drafted.

7. Research and development endeavours, particularly the drawing up of a register of current and completed research and de-

velopment projects run by the Member States and the Community,
and identification of the research and development efforts
still needed in this high-risk sector.

The second part of the medium-term waste programme "other
problems" covers the following points :

1. Possible additions to the list of substances in the Annex
   to the Directive.

2. An exchange of information and experience about central
   treatment plants.

3. Packaging and conditioning (selection, criteria and regulations for containers, decontamination, disposal of used
   containers, etc.).

4. Insurance and liability.

5. Recycling and re-utilization as secondary raw materials.

6. Replacement of materials and processes, particularly to
   reduce the production of toxic and dangerous waste.

7. Incineration and generation of energy.

8. Chemical, physical and biological treatment methods.

9. Waste which does not fall until now within the scope of
   the Directive.

As regards such waste not covered by the Directive, e.g., hospital
waste, certain types of mining waste, etc., the idea is being
considered of either drafting separate Directives or of harmoni-
zing the principles of existing or planned laws and administrative
provisions of the Member States in this field by means of framework Directives, recommendations, guidelines and/or codes of
practice.

The need for uniform identification and accompanying documents,
in particular for transfrontier transport of this waste, arises
from the fact that to an increasing extent toxic and dangerous
waste is being carried across the national frontiers of the
Member States. At present, control over toxic and dangerous waste
ends at the national frontiers even of those countries which already have an elaborate control system. From the point of view of
the country of dispatch this waste disappears into outer darkness
once it has crossed the frontier. However, within the unified
economic area of the European Community control over this highly

dangerous waste must be continuous from first production to final disposal. There have been many incidents in the past which showed that the transfrontier carriage of toxic and dangerous waste is a particular source of danger to human health and the environment. For this reason the Commission takes the view that there must be standard identification and accompanying documents in several languages for the transfrontier carriage of such waste.

This would at the same time facilitate transfrontier carriage which - because of highly divergent national regulations - is subject to a highly bureaucratic and cumbersome procedure.

Another important problem area, in respect of which the Directive of 20 March 1978 needs to be supplemented, is the <u>transport</u> of this waste. Since these substances are to an increasing extent being carried over great distances, there are particularly high risks involved, risks which are in many cases greater than those involved in the temporary and final dumping or treatment of this waste.

Although there are international conventions on the carriage of dangerous goods, these have still not been ratified and put into effect by all the Member States. Furthermore, they apply only to international and not domestic transport. Some Member States have applied the rules of these international conventions to domestic transport as well. However, not all Member States have done so, in fact some have adopted quite different regulations of their own, which differ from the international conventions on international transport. Other Member States again have no specific regulations at all relating to the carriage of dangerous goods. In other words, this matter is not regulated uniformly within the Community. In view of the particular risks in this sector and the many major accidents which have already occurred, a uniform system is urgently necessary within the Community.

The Commission is not necessarily aiming at a separate Directive on the transport of toxic and dangerous waste. The matter could, for example, be dealt with by extending existing regulations on the carriage of dangerous goods to those substances and products which they so far fail to cover. The main thing is that there should be a uniform Community system covering the carriage of dangerous goods and that it should satisfactorily settle the specific problems of toxic and dangerous waste. These problems arise mainly from the fact that toxic and dangerous waste often involves compounds that are not to be found among the substances covered by the regulations on the carriage of dangerous goods. In other words, we have here a borderline area within the carriage of dangerous goods sector which is in need of regulation.

Another important problem with which the Community has to concern
itself is the matter of <u>concentrations</u> of toxic and dangerous waste.
The Directive of 20 March 1978 contains no concentration values.
However, when the Directive was adopted by the Council, the Commission undertook to make a thorough investigation of this question
after the Directive was adopted.

At present <u>all</u> the substances listed in the Annex to the Directive
are subject to the provisions of that Directive as regards, for
example, authorization procedure, accompanying papers, control, etc.,
regardless of the concentration in which they occur. Article 1 of
the Directive defines toxic and dangerous waste as follows:

" 'toxic and dangerous waste' means any waste containing or contaminated by the substances or materials listed in the Annex to this
Directive of such a nature, in such quantities or in such concentration as to constitute a risk to health or the environment."

Within the Community three Member States - Belgium, the Netherlands
and the United Kingdom - have set concentration values at a national
level. One other Member State is at present drafting regulations
setting out concentration values.

There is evidently no intention at present in the other Member
States of setting concentration values for substances or materials
contained in toxic and dangerous waste.

Regulations already existing or being drafted differ widely from
one Member State to another in respect of concentration values,
criteria, scope, application and the like. In some countries
concentration values are generally applied, while in others they
apply only to specific disposal methods, transport and storage.

The Commission must see to it that the divergences between the
Member States as regards levels of and criteria for concentration
values do not impair full application of the Directive of 20 March
1978 or the proper working of the common market.

That is why it is urgently necessary - quite apart from the obligation assumed by the Commission - that all the problems related
to toxic and dangerous waste be examined in relation to the determination of concentration values for the substances and materials
listed in the Directive.

The Commission has therefore set up a working party of scientific
and technical experts on toxic and dangerous waste. This working
party has been given the following tasks:

- to examine the basic problems, particularly in respect of the
  suitability of concentration values, basic criteria, scope of
  application etc;

- to collate and assess existing technical and scientific knowledge
  and available experience;

- to examine which concentration values might be proposed at
  Community level;

- to examine what effects - if any - on the working of the common
  market and the competitiveness of undertakings result from
  divergent national concentration values.

This working party will have to submit a report with appropriate
proposals to the Commission.

The proper management of toxic and dangerous waste requires
suitable treatment plants and special dumping sites. Article 12
of the Directive therefore prescribes that plans be drawn up
covering the necessary special treatment plants and suitable
disposal sites. This stipulation logically supplements the
requirement that toxic and dangerous waste may be treated or
disposed of only in authorized plants.

Because of the special dangers to man and the environment and
in order to prevent water pollution and provide protection against
nuisances, particular demands have to be made on sites and dumps
where toxic and dangerous waste is temporarily or finally deposited. The selection, supervision and long-term safeguarding of
sites and dumps for toxic and dangerous waste are therefore among
the priorities of environmental policy and waste management.

A number of Community countries, the Netherlands in particular,
have no - or insufficient - sites with the required features.
This means that dumps and treatment plants must be accessible for
waste from other Community countries as well.

Toxic and dangerous waste often occurs in small and widely dispersed quantities, which means that, for reasons of economic
efficiency, central facilities with an appropriate number of
collecting points and temporary dumps are needed. In the interest
of economical, safe and environmentally sound treatment and
disposal of such waste, the areas served by such central facilities within the Community have to be supraregional and international.

This is why - for practical, economic and ecological reasons -
there is transfrontier carriage of toxic and dangerous waste,
which is constantly increasing in both extent and frequency.

Despite the Directive of 20 March 1978 there are no uniform
regulations in respect of problems relating to the transfrontier
carriage of toxic and dangerous waste, though such regulations
are needed, since the lack of them increases the risks involved

and has already given rise to serious hazards.

This situation provides a number of reasons why the Directive
78/319.EEC of 20 March 1978 must be supplemented and given
substance. In particular, priority community rules and provisions
are needed in the following areas:

- firstly, the obligation to give notification of cross-frontier
  shipments of toxic and dangerous wastes;

- the use of a standard accompanying form made out in the official
  Community languages for cross-frontier shipments. At present,
  the inspection and notification procedures stop at national
  frontiers;

- uniform rules concerning safety instructions relating to
  hazards and accidents during transport, particularly in the
  case of cross-frontier shipments;

- uniform rules for the import and export of toxic and dangerous
  wastes; some Member States have special legislation in this
  area which also applies to intra-Community trade; others have
  only general regulations - particularly with respect to non-
  member countries - or no regulations at all;

- uniform rules for recipients, the means of transport, pretreatment and manner of packing toxic and dangerous wastes
  (selection and criteria; regulations governing the recipients,
  decontamination, disposal of used recipients and containers,
  etc.);

- authorization also for the carriage of toxic and dangerous
  wastes;

- uniform rules governing the selection, monitoring and management of storage and disposal sites for toxic and dangerous
  wastes;

- rules governing access to treatment centres and dumps in other
  Member States of the Community;

- provisions relating to civil liability and insurance. The
  increasing number of disposal sites for toxic and dangerous
  wastes and the growing need to consult the public on this
  matter are making it more and more urgent to clarify the
  problems of civil liability and insurance;

- preparation of a Directive on the disposal of toxic and
  dangerous wastes at sea;

- consideration of whether there is a need to define the concentrations of the substances listed in the Annex to the Directive
  and, if so, what the concentrations should be; proposals could
  then be submitted to the Council;

- formulation of a framework Directive on the disposal of hospital wastes which were deliberately excluded from the scope of the Directive with a view to prepare a special Directive. Some hospital wastes contain pathogenic substances which are very dangerous. Government experts, the European Parliament and the Economic and Social Committee have stressed the need to supplement Directive 78/319/EEC by a special Directive dealing with hospital wastes with a view to establishing common rules in particular for the pathogenic substances concerned.

The legislative programme which must be undertaken in the field of toxic and dangerous waste over the next two years covers the following three areas :

- full and rapid implementation of the provisions of Directive 78/319/EEC;

- harmonization of the national implementing laws and other existing national regulations in this field since there are often marked differences between them;

- drafting of Community implementing regulations to supplement Directive 78/319/EEC.

The problems of civil liability, insurance and financial guarantees are among the most significant as regards the management of toxic and dangerous wastes in view of their special nature.

Liability provisions are needed not only for emergencies but also for damage caused, e.g., by the migration of toxic substances from dumps, and for the long-term safety of installations and dumps which have already been closed. In the latter case, liability could extend over several decades following the cessation of economic activity.

If the owner does not have the financial means to compensate victims, the national authorities may be obliged to bear the costs.

One must therefore clearly ascertain where liability falls and adequate insurance provision must be made, including if necessary the creation of funds to cover long-term underwriting.

A number of countries have already made special provisions in their legislation, but others continue to determine civil liability within the traditional framework of civil law.

An exchange of information and experience on this topic not only within the Community but also on a large international scale seems to us extremely important and urgent.

Having in mind the particular potential risks involved, it should
be considered if objective liability would not be the appropriate
solution.

Another key problem is the relationship between dumping and the
physico-chemical treatment of toxic and dangerous wastes. This
latter method of disposal is generally regarded as too expensive;
it is for this reason that some four or five plants have gone
bankrupt in the last two years for want of sufficient quantities
of waste to handle. As a result of financial constraints, there
is a growing trend towards dumping and towards searching for
suitable new disposal sites, since current demand is outstripping
the capacity of the existing tips and dumps. Other treatment
centres are likely to be forced to close; this will lead to a
proliferation of "artificial" dumps. Nevertheless, adequate
capacity for physico-chemical treatment must be maintained and
any proliferation in the number of tips for toxic and dangerous
waste prevented. Community action is certainly required to
promote cross-frontier cooperation (large-scale treatment centres)
and to draw up stringent criteria for dumps authorized to receive
toxic and dangerous wastes.

Toxic and dangerous waste derives mainly from chemical manufacturing processes. In other words, they are closely related to
industrial activities. The costs involved in waste are increasingly important. Therefore uniform solutions are necessary within
the community, so as to avoid market disruptions and distortions
of competition resulting from divergent or even contradictory
national regulations. Another economic or industrial consideration
is that a whole series of toxic and dangerous wastes, which are at
present destroyed because technological development is still
inadequate, are highly valuable potential raw materials. Examples
are the halogenated hydrocarbons, used solvents and the sludge
from galvanizing processes.

Thought should therefore be given to the question whether continued destruction, which involves heavy costs, might not be
replaced by temporary storage in the common interest of the
community countries.

All these considerations make research and development activities
particularly important, and the Community should therefore set up
a specific research programme on toxic and dangerous waste as part
of its research policy. Such a programme should aim at acquiring
detailed chemical and physical knowledge and experience about the
long-term effect of such substances and compounds, as well as
providing knowhow which will reduce the quantity of and the
dangers inherent in this waste and result in increased recycling.
In almost all the Member States and in the major industrialized

countries long-term research is underway. There should however be far greater circulation of information regarding the subjects and results of research and testing activities. The setting up of a special data bank on toxic and dangerous waste therefore seems both useful and necessary. There should, in addition, be a regular exchange of experience and views on major topical or basic questions.

In order to make R&D actions more efficient, the Commission is considering the following activities and procedures :

- establishment and regular updating of an inventory or register of the research and development work being carried out on toxic and dangerous waste in the Community and in non-member countries;

- identification of the research and development work still needed in this special risk area;

- formulation of a Community R&D programme in this field or of Community R&D measures to be carried out under the environmental research programme;

- development of a procedure for publishing the research findings in individual Member States and in the Community as a whole;

- organization of meetings to exchange information and experience on current research and the findings made.

Frank, trusting and close cooperation between industry and the public authorities is of decisive importance to the success of a management policy for toxic and dangerous waste. This is for several reasons:

In the first place, such waste derives mainly from industry.

In the second place, treatment and disposal of such waste requires a highly developed technology.

Thirdly, the Commission takes the view in the light of the "polluter pays" principle accepted by the Community and of the liberal principles on which the Community is based, it is primarily industry's task to solve the problem related to the production, treatment and disposal of toxic and dangerous waste within the framework of conditions set by the authorities.

We do not, therefore, believe that the disposal, the collection, the treatment and the dumping of dangerous waste should be a monopoly of the public authorities. This should be a normal sector of our economic system, as long as industry is capable of coping with the problems and the requirements of protecting the environment are met.

The task of the authorities is to exercise stricter control and supervision in this particularly hazardous field and to establish the conditions required for safety by means of legal and administrative measures.

In the work done so far in the field of toxic and dangerous waste there has always been close cooperation between the Commission and industry, which made its knowhow and experience available to us in the drafting of the Directive and in the follow up.

It is our intention to continue this practice also in the second phase of our endeavours. Industry has already provided us with valuable recommendations for guidelines for the treatment and disposal of various hazardous substances.

In view of the particular potential hazards of this waste and its complex nature, highest priority will at all events also in future be accorded to toxic and dangerous waste under the Community's waste management policy.

ASPECTS OF THE HAZARDOUS WASTE PROGRAMME IN THE NETHERLANDS

A. Goudsmit
Ministry of Health and Environmental Protection

F. van Veen
Infra Consult

C.J. Duyverman
Organization for Applied Scientific Research TNO

THE NETHERLANDS

INTRODUCTION

The disposal of wastes of all kinds in an environmentally acceptable manner presents a complex problem in the Netherlands. Densely populated and highly industrialised, with an extensive use of its soil for agriculture and dependent for its water supply on ground and surface water, the country, a large part of which is low-lying, offers but few areas which are suitable for waste disposal by landfill. The scarcity of landfill locations and at the same time the necessity to save raw materials and energy, have determined the waste disposal policy and, consequently, the waste disposal programme of the Netherlands government. Thus, the generation of domestic and industrial waste in general, and of hazardous waste in particular, has to be reduced wherever and whenever possible. Processing and re-use of wastes and recovery of materials and energy from wastes have accordingly to be maximised so as to reduce to an absolute minimum the amount of waste which can only be disposed of by landfill.

This approach to waste disposal applies in particular to hazardous waste which in the Netherlands is regulated by the Chemical Waste Act as the principal legal instrument. The disposal of other categories of waste is regulated by the Waste Act, whilst separate legislation applies to radioactive waste. The Chemical Waste Act which was implemented in 1979 is aimed at reducing and finally preventing the pollution of the environment by chemical waste and used oil. The substances to which the Act applies are listed in the Schedules to the Substances and Processes Decree.

137

The list was established on the basis of such aspects as toxicity, including cumulative effects, persistence and possible harmful effects on either man, animals, plants or the biosphere as a whole.

The provisions of the Chemical Waste Act - which do not apply to private households - compel waste generators wanting to dispose of their chemical waste and used oil by handing these wastes over to another person, to report the relevant details - composition, properties, quantity and destination - to the Minister of Health and Environmental Protection. The storing, treating, processing and destruction of chemical waste and used oil are subject to a licence granted by the minister; only licensed people may deal with waste which others have disposed of and then only when a description of that waste has been provided. There is a comprehensive duty on licensees to report details of the transaction to the minister. Furthermore the Act prohibits the disposal of chemical waste and used oil by depositing these wastes in or on the soil, whether in a container or otherwise. Exemptions from this prohibition may be granted by the minister but only when alternatives (e.g., treatment, intermediate storage, etc.) are not available, and only under strictly controlled conditions.

RESEARCH AND DEVELOPMENT ACTIVITIES

In the Netherlands chemical waste is generated by many industrial activities in large quantities covering a wide range of substances. Increasing amounts of chemical waste are produced as a result of the introduction of processes, enforced by environmental legislation, to treat waste water and flue gases. The high costs of waste disposal, on-site as well as off-site, the problems encountered by waste generators in finding suitable landfill sites and obtaining exemptions from the prohibition order have led many industries to reduce the amount of their waste by application of "good housekeeping", by changing their processes and by replacing raw materials. For a decade, the Netherlands' government has stimulated the research and development of processes which can prevent the generation of hazardous waste and their discharge into the environment. For this purpose financial grants are available for universities, research institutes and industrial research establishments.

As an example of the projects aimed at replacing hazardous metals, the replacement of cadmium in metal plating can be mentioned. Replacing cadmium by an other metal has proved to be possible in several cases, e.g., in solder. However, for applications in which the product has to meet a high performance level, cadmium is still required. A project has recently been started to investigate whether in surface layers cadmium can be substituted by aluminium. To this end aluminium is precipitated from a solution and the layer thus obtained has a purity of more than 99,99%. Its properties equal those of cadmium layers and further evaluation for various industrial applications is in progress.

When replacement of hazardous metals is not possible, a modification of the process used may be more succesful, such as the installation of spare baths and the making of special provisions to prevent metals from contaminating rinse waters. An example of how the metal concentration in rinse water and thus pollution can be prevented is the electrochemical precipitation of nickel and/or zinc from rinse baths. This technology, well-known for cadmium as well as for precious metals, may under certain conditions be also applicable to nickel and zinc. The Netherlands' consultancy bureau Infra Consult together with the Organisation for Applied Scientific Research TNO are currently investigating this possibility by using two-dimensional as well as three-dimensional electrodes in order to find the optimum conditions.

The two above-mentioned investigations represent typical examples of industrial research projects which are currently in pro-gress in the Netherlands and which aim at replacement of hazardous metals. In addition a wide range of projects is aimed at the processing of hazardous waste. In view of the environmental hazards of the landfilling of metal containing chemical waste it is not surprising that in the Netherlands with its stringent landfill policy, the hazardous waste program is mainly directed to the treatment of such wastes, in particular of mixed metal wastes. In general the treatment of mixed metal waste consists of the extraction of some of the metals, in which case a less contaminated residue is produced, or in the total extraction of all metals in which case the residue obtained is inert. Four major projects on mixed metal wastes are currently in progress. The processes under investigation have the same object viz. to recover the metals and to prevent the waste from being dumped; they are based on different principles:
- selective anion-exchange
- chloride metallurgy
- solvent extraction
- physical separation and leaching
Short descriptions of these processes are given in the next chapter.

RECOVERY PROCESSES FOR MIXED METAL WASTES

1.   TNO-process for metal recovery based on anion-exchange
      (figure 1)

Metal hydroxide sludges, concentrated liquids and comparable wastes containing Fe, Ni, Cr, Cu, Zn, Mn, Co, and Cd can be treated so that pure metal salts of all these metals are recovered and a residue is produced which is only slightly contaminated (on a ppm level) and which can be deposited without any potential danger to the environment. The process is based on the formation of metal chloride complexes with concentrated hydrochloric acid. After filtration of the inert residue, the metals are separated from the solution. First Fe is removed by solvent extraction with methylisobutylketone. The metal complexes are subsequently passed on to an anion-exchanger where they are bound to the exchanger material. Separation of the metal choride complexes, followed by eluation of various metalchlorides is performed by applying solutions of decreasing concentration of hydrochloric acid. In the eluting step the complexes decompose to the chlorides.

Nickel and chromium are not complexed under the process conditions and are thus not fixed to the anion exchanger. Chromium and nickel are separated after Cr-III has been oxidised to Cr-VI.

The process has been investigated on a laboratory-and on pilot-plant scale (capacity 100 kg per day). It shows separation to be possible, yielding pure metal salts and an inert residue.

An estimate, based on fixed and variable costs, shows costs to be in the order of ƒ 250,- per ton of waste for a plant capacity of 10.000 tons per annum.

The process has a broader field of application than for the treatment of hydroxide sludges alone. In principle any material containing metals which can be transformed to chlorides, can be separated into the individual metals along the lines of this process. In Appendix 1 a more detailed description of the process is given.

2.   Chloride metallurgy process for the recovery of metals
      (figure 2)

The application of chloride metallurgy has so far been restricted to the enrichment of lean ores and the refining of certain metals. The process is based on the conversion of metal oxides to chlorides with hydrogenchloride (HCl) in the absence of oxygen and at temperatures of 400-1000°C. This process will be investigated for car-shredder waste, initially on a laboratory scale. This waste is composed of plastics, such as polyvinylchloride, rubber, paint, iron and other metals such as Cd, Cu, Pb, Zn and Ni. Pyrolysis of this waste produces volatile hydrocarbons and

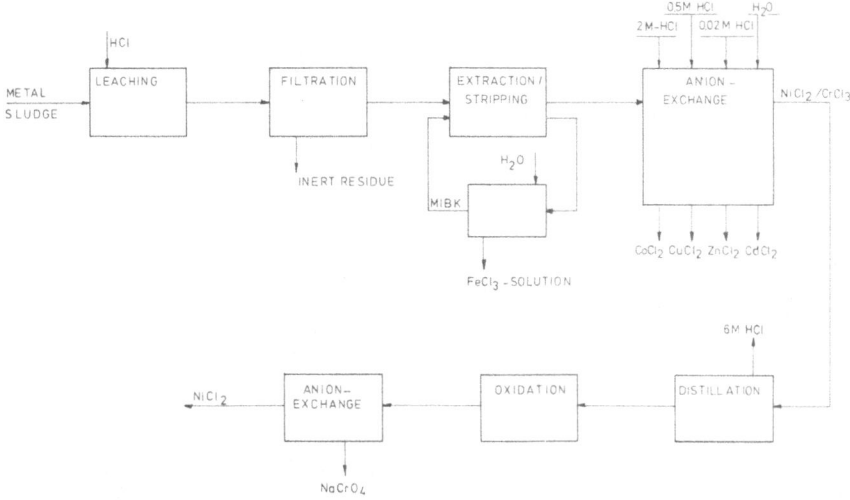

Figure 1.   Recovery of metals based on Anion-Exchange (TNO Process).

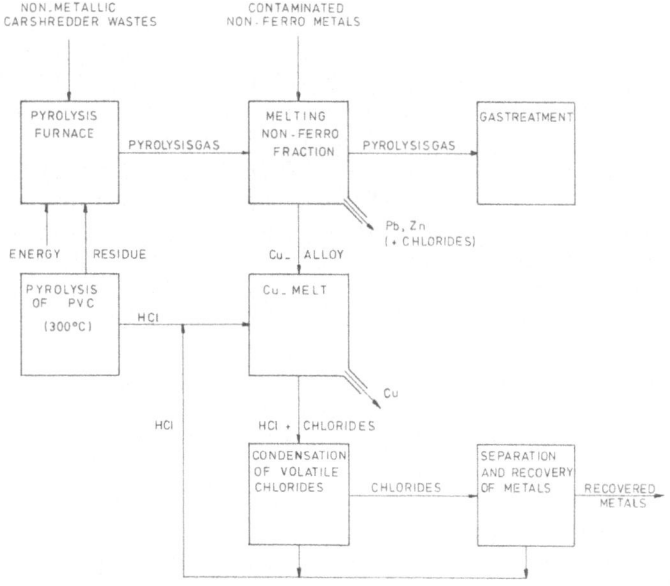

Figure 2.   Recovery of metals based on Chloride-Metallurgy.

hydrogenchloride. PVC waste can be pyrolysed separately to produre sufficient HCl-gas. The volatile hydrocarbons can be applied as fuel and for the reduction of chlorides and oxides.

Furthermore, the non-ferro fraction of the car scrap containing mainly Zn, Pb and Cu, is heated to a temperature of about 500°C: Zn and Pb melt and can be separated according to usual metallurgical practices. The excess of HCl reacts with Zn to $ZnCl_2$ and can be removed at the Cu-refining stage. The volatile metal chlorides are condensed. Investigations into the separation of these products are part of the research activities carried out by the Technical University Twente.

In Appendix 2 a more detailed description of the application of chloride metallurgy to waste materials is given.

3.   Process for the recycling of pure metal salts, based on solvent extraction

The Technical University Eindhoven has developed a process for recycling the metal salts of Cr, Cu, Ni, Cd and Zn from mixed metal hydroxide sludges. The process, still on a bench scale, consists of three steps: In the first step Cr (III) is oxidized to Cr (VI). This can be done in several ways. Investigations to select the most economical method are still in progress. After separation of the chromate solution the remaining sludge is leached in the second step with diluted sulphuric acid at a pH of 3. At this pH Fe (III) remains unchanged in the sludge while the other metal ions are dissolved. It is not economically attractive to recover Fe (III). The aqueous solution obtained through the leaching step contains the metal ions Cu, Zn, Cd and Ni. In the third process step these metal ions are separated in several liquid-liquid extraction stages. All extractions are based on the application at different pH-values of one single extraction agent. This agent, which was especially developed for this purpose at THE, is a mixture of cyclohexanone oxime and oleic acid (Dutch patent nr.77 02517). Any usual organic solvent may be used. After one extraction step the extracted metal ion is stripped from the organic phase with a sulphuric acid solution. The stripping solution is recycled until it is saturated with the metal sulphate. The metal sulphate will then crystallize, which serves as an additional means to purify the salt.

4.   Process for the treatment of blast furnace waste (figure 3)

A good example of solving a landfill problem is the method investigated at the iron and steel plant of ESTEL HOOGOVENS B.V. at IJMUIDEN.

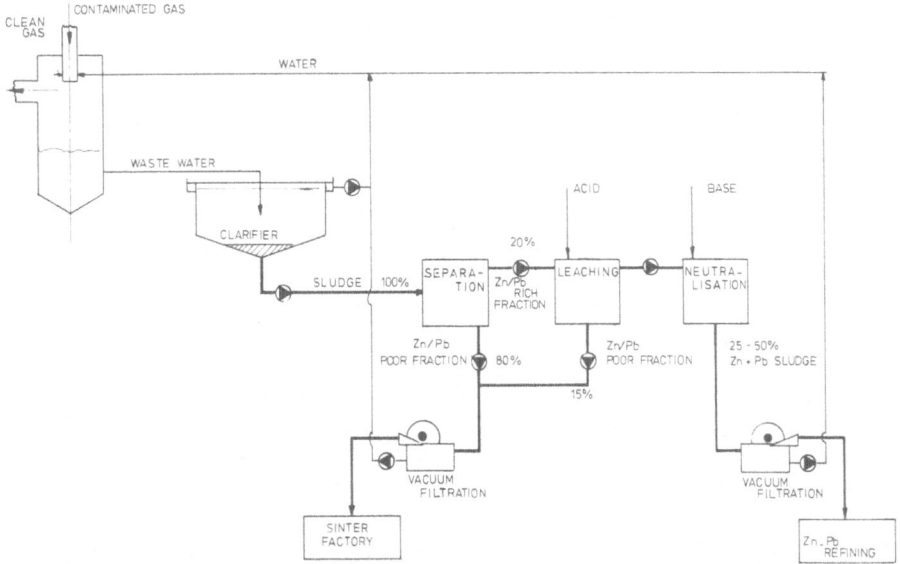

Figure 3.   Blast Furnaces, Gas Cleaning and Sludge Treatment.

Blast furnaces produce an off-gas containing zinc and lead as the principal metallic contaminants. The gases are wet-scrubbed. Subsequently the waste water is clarified, producing a metal bearing sludge. The sludge is treated in such a manner that a Zn/Pb-rich and a Zn/Pb-poor fraction are obtained. The poor fraction can be recycled to the sintering plant, while the rich fraction is leached resulting in a concentrated liquid rich in Zn and Pb. After the metals have been precipitated from the liquid, a sludge is obtained which is further processed for the ultimate recovery of the metals.

This project shows very good possibilities for economic application, at least when it is compared with the rather high costs of alternative methods of waste treatment, e.g., pyrometallurgical techniques.

CONCLUDING REMARKS

In conclusion it has to be emphasized that the above-mentioned processes are still in the bench- or pilot-plant phase. Further investigations and economical evaluations have to show whether the various processes will be feasible on a commercial scale. The economic feasibility, in particular, will be determined by the scale of the operation. In this respect the chances of success may be considerably enhanced when sufficient amounts of the waste to be processed are available from a small number of sources such as in the case of the process for the treatment of blast furnace waste described above (in fact a single source in the Netherlands). The economics will be adversely affected, however, when the amount of waste required for commercialisation of the process can only be obtained from a large number of sources, each offering relatively small amounts. Such a situation appears to exist in the case of the above-described anion-exchange process for which the metal containing waste originates from several hundred sources. The added costs of transport to a central processing facility as well as the costs of preparing a suitable feedstock from many batches of metal containing waste of often widely varying composition may play a dominant role in this case. In addition the problem arises that in the Netherlands the quantities of suitable mixed metal waste available may not add up to the minimum plant capacity considered to be required for a succesful technical realisation of the process. This situation is unlikely to be unique to the Netherlands nor does it appear to be limited to mixed metal wastes only. Co-operation with other countries on the realisation of processing plants for specific types of hazardous waste of a sufficiently large size - also to compensate for the added costs of transport etc. - may well prove to be the only

alternative to each country operating plants with prohibitively
high costs to the users or to such plants not becoming available
at all.

ACKNOWLEDGEMENTS

The authors gratefully acknowledge Prof.Ir. M. Tels and
Ir. A. Bolt of the Technical University Eindhoven,
Prof. Dr. Ir. B.H. Kolster of the Technical University Twente and
Mr. W. Kat of ESTEL HOOGOVENS B.V. for their contribution to this
paper.

Appendix 1

SHORT DESCRIPTION OF TNO'S PROCESS FOR THE RECOVERY OF METALS FROM
METAL FINISHING SLUDGE

INTRODUCTION

In the metal finishing industry, waste water treatment usually
comprises detoxification of waste waters containing metal cyanides
and/or chromium-VI, followed by neutralisation and precipitation
of heavy metals with sodium hydroxide or lime. Sludges of heavy
metal hydroxides are thus obtained. These sludges contain the
following metals: iron, nickel, chromium, copper, zinc, manganese,
cobalt, lead, cadmium and aluminium. The heavy metals having toxic
properties, disposal of the sludges presents environmental
problems. In the Netherlands, the Chemical Waste Act prohibits
dumping of these sludges. In view of these environmental problems
and because recovery may be an attractive solution to these
problems, the Ministry of Health and Environmental Protection has
granted financial aid to the Organization for Applied Scientific
Research for the development of a process to recover the heavy
metals from the sludges. The residue of such a process must not be
toxic, so that it can be disposed of as a non-hazardous waste.

The main features of the process under development

The sludges are dissolved in hydrochloric acid, in which
nearly all metals form negatively charged chloride complexes.
These complexes are bound to a strongly basic anion exchange resin.
By subsequently eluting the anion exchanger with hydrochloric acid
solutions of decreasing strength a separation of the metals is
achieved, yielding pure metal chloride solutions.

Up to now experiments on a bench- as well as on a pilot-plant
scale have been carried out successfully. On the basis of the
results of these experiments, an engineering flow scheme for a
plant with a capacity of 10.000 tons of sludge per year has been
drawn up, and a cost estimate for the whole process has been made.

Application of the process is not restricted to metal
hydroxide sludges. In principle the scope of this chromatographic
separation process is a much broader one; for example waste
materials containing metals can be treated for the recovery of
pure metals or metal salts, if the metals can be transformed to
the chlorides.

THE PROCESS

The recovery process consists of the following steps: (see figure 2)
- dissolving the sludge
- filtration of insoluble material
- extracting and stripping the iron compounds
- anion exchange and elution of the metal chloride solutions
- separating nickel and chromium
- concentrating the zinc- and/or cadmium chloride solutions.

Dissolving the sludge

Dissolving the sludge is achieved by adding concentrated liquid or gaseous hydrochloric acid. The following reactions take place:

$$Me(OH)_x + x\ HCl \qquad\qquad MeCl_x + x\ H_2O \qquad\qquad (a)$$

by which the metal hydroxides are converted to the corresponding metal chlorides. By adding excess HCl most of the metal chlorides are bound to complex compounds in accordance with:

$$MeCl_x + Cl^- \qquad\qquad MeCl^-_{(x+1)} \qquad\qquad (b)$$

In this manner cobalt, manganese, iron, lead, zinc and cadmium are complexed. Nickel and chromium remain uncomplexed.

Filtration

The sludge contains different insoluble materials, mainly silicates, orginating from degreasing baths. To avoid these compounds interfering in subsequent processing, the insolubles have to be removed by filtration.
Thorough washing of the filter cake with a hydrochloric acid solution is necessary to remove all the metals from the cake, so that its metal content meets the requirements set by the Chemical Waste Act. The washing liquid is returned to the dissolving vessel.

Extracting and stripping the iron

As mentioned above, iron is complexed to $FeCl_4^-$, in accordance with reaction (b) and it can therefore be separated from the other materials on the anion-exchange column. However, because iron is normally present in the sludge in relatively high concentrations, it would charge a large section of the (rather expensive) anion-exchanger.

Iron being a relatively inexpensive metal, the costs for separating it by means of the anion-exchanger would not be made up by its sales revenues.
Therefore iron is removed from the metal-bearing solution by extraction with methylisobutylketone (MIBK). In a one-stage extraction process nearly 99.9% of the iron is extracted; by stripping the iron-loaded MIBK with water an iron chloride solution is obtained. The MIBK is returned to the extraction vessel.

Anion-exchange and elution of the metal chloride solutions

The iron-free solution, which is 6 molar in HCl, is passed through a column filled with a strongly basic anion-exchange resin. Here the metal complexes are bound, nickel and chromium pass through and are collected together at the bottom of the column. Next the anion-exchanger is eluted with hydrochloric acid solutions of decreasing strength in order to recover the metals separately. These separating steps are based on the different complex constants of the individual metal complexes. By lowering the Cl^- - concentration (i.e., the strength of the HCl-solution) reaction (b) is shifted to the left, which means that the various complexes are decomposed depending on the Cl^--concentration and the metal ions are released by the anion-exchanger. In 2 molar HCl for example the cobalt complex decomposes and the cobalt ion is eluted; copper, zinc and cadmium remain complexed and remain bound to the ion-exchanger. By repeating this procedure with HCl of decreasing molarity these metals are separated and eluted from the column in successive steps.

Separating nickel and chromium

The nickel-chromium solution that passes unretarded through the column is 6 molar in HCl. Before any subsequent treatment of this mixture can be carried out, the HCl has to be removed. This is done by distillation, producing a 6 molar HCl-azeotrope, which can be recycled to the dissolving step. At a pH of 4, Cr-III in the nickel-chromium mixture is then oxidised to Cr-VI with sodium hypochlorite. Two methods can be applied to separate nickel and chromium:
-    Cation-exchange; by this method nickel is bound, chromium passes through;
-    Anion-exchange; chromium is bound and nickel is collected.
The anion-exchange route has been investigated. Cr-VI (as chromate) is bound to a strongly basic anion-exchanger and subsequently removed by sodium hydroxide.

Concentrating the zinc and/or cadmium solutions

The concentrations of zinc and cadmium in their respective solutions as eluted from the column are rather low (5 and 1 g/l). By passing them through cation-exchangers and subsequently eluting the latter with an aqueous HCl- or H_2SO_4-solution concentrated chloride or sulphate solutions containing 50 - 60 g/l of metal are obtained.

EXPERIMENTAL PART

The pilot plant consists of:
- dissolution vessel: contents 0.5 m^3
- filtration unit: surface area 1.5 m^2
- extraction vessel: contents 0.5 m^3
- anion-exchange column: length 2.2 m, diameter 0.4 m,
  bed height 0.55 m, exchange
  capacity about 85 equivalents.
- distillation unit
- several dosing systems for gaseous and aqueous
  hydrochloric acid

For the separation of the Ni/Cr-solution additional equipment is used, the main components of which are an oxidation vessel and an anion-exchange column. The capacity of the installation amounts to about 100 kg sludge per day.

RESULTS OF TECHNICAL EXPERIMENTS

About ten different sludges have been treated on a bench scale as well as on a pilot-plant scale. It could be shown that metals can be sharply separated; less than 10 ppm cross-contamination can be achieved.
The metal contents of the filter residues are very low in all cases and do not in any case reach the level which would classify them as chemical waste.

Though manganese and lead are usually present in minor amounts, they may in certain cases be present in higher quantities, in which case they can seriously interfere with the separation process. Since manganese is collected together with nickel and chromium, a process for separating nickel from manganese has to be used as a final step. Lead gives rise to problems arising from the low solubility of lead chloride, which causes it to precipitate in the anion-exchange column. Further investigations have to be carried out to solve this problem.

An example of the separation of metals on a pilot-plant scale from a mixed sludge of various metal-finishing processes is shown in Table 1. The separation has not been optimised in this case. Nevertheless Table 1 shows the potentialities of the process.

RESULTS OF ECONOMIC EVALUATIONS

On the basis of the results of the experiments a preliminary economic evaluation has been made, based on the following assumptions:
- Plant capacity   : 10.000 tons sludge/year
- Depreciation time   : 5 years
- Lifetime ion-exchange resin: 3 years
- Operation time   : 16 hours/day, 300 days/year

Selling price of the different metal salts has been fixed at 70% of the market price of these salts.
Investment costs: Dfl. 5 x 10^6 (about $ 1.8 x 10^6)
Nett costs (fixed plus variable costs minus yield of metal salts): Dfl. 250,-/ton sludge (about $ 90,-).

TABLE 1: Metal separation from a metal hydroxide sludge

- The sludge contains Ni, Cr, Mn, Co, Cu, Fe, Pb, Zn, Cd and Ag.

- Composition of the filter residue after slurrying the sludge, treating the slurry with gaseous hydrochloric acid, filtering the slurry and washing the residue: metal content residue < 0.05 % (based on dry filter cake)

- Composition of the methylisobutylketone phase ion (g/l)
  Ni: 0.007; Cr: 0.19; Mn: 0.000; Co: 0.06; Cu: 0.13; Fe: 18.8; Zn: 0.8; Cd: 0.1;

- Composition of the aqueous phase after extracting iron with methylisobutylketone (g/l):
  Ni: 66.2; Cr: 23.6; Mn: 0.15; Co: 2.4; Cu: 4.3; Fe: 0.007; Pb: 0.54; Zn: 4.8; Cd: 0.85; Ag: 0.62.

- Composition of the various fractions after eluting the anion-exchange column:

Fraction: Composition (g/l)

		Ni	Cr	Mn	Co	Cu	Fe	Pb	Zn	Cd	Ag
1.	(Ni,Cr,Mn)	3.5	8.4	0.07	0.004	0.000	0.001	0.11	0.000	0.000	0.01
2.	(Co)	0.04	0,03	0.0002	4.3	0.001	0.0006	0.12	0.000	0.000	0.04
3.	(Cu)	0.02	0.02	0.030	0.01	6.7	0.01	0.2	0.06	0.000	0.02
4.	(Zn)	0.005	0.002	0.000	0.0008	0.008	0.0001	0.03	1.26	0.0007	0.00
5.	(Cd)	0.002	0.0005	0.000	0.000	0.002	0.0001	0.00	0.005	0.15	0.00

- The separation of nickel and chromium in fraction 1 was successfully carried through on a bench scale.

Appendix 2

INVESTIGATION INTO THE POSSIBILITIES OF APPLICATION OF CHLORIDE METALLURGY TO WASTE MATERIALS AND COMPLEX METAL WASTES

Chloride metallurgy already has been applied for a long time for the enrichment of lean ores and for the refinement of metals. It is based on the conversion of metal oxides to chlorides with the aid of HCl or Cl_2 in the absence of oxygen in a temperature range of 400 - 1000°C. For this purpose use is made of the differences in thermodynamic stability and vapour pressure of the chlorides formed.

At the Technical University Twente research has been conducted into the possibilities of the separation of metals from surface layers, such as the separation of tin from tinplate, zinc from galvanised iron, tin from copper, chromium from copper, etc. To this end use was made of polyvinylchloride (PVC), from which by subjecting it to an individual pyrolysis process at 300°C nearly pure hydrogenchloride is liberated. This gas is passed over the metal, which has previously been raised to the temperature required for the separating process. In several experiments the PVC was pyrolysed together with the metal at the same temperature. As a result of the higher temperature during the latter process not only HCl, but also volatile hydrocarbons are released[*].

For separating tin from tinplate a search has been made into the conditions required for temperature and residence time for the treatment of small amounts of tinplate on a laboratory scale. The results of the pyrolysis experiments show that the process is eminently suitable for the detinning of tinplate separated from domestic waste and from the ash residues of waste-combustion installations. Tin percentages lower than 0.05% can be achieved. Even the tin from the soldered seams can almost completely be removed in this way. A temperature of approx. 600°C will do for a rapid tin extraction from tinplate separated from domestic waste (10-15 minutes). The loss of iron is minimal with this process. For tinplate obtained from waste incinerating plants it is recommended to use temperatures over 700°C to remove the tin that has diffused into the plate material. The loss of iron will slightly increase in this case. Lacquer and paint layers and adhering dirt are adequately degraded in the pyrolysis process. The minimum amount of HCl required for the conversion of tin and tin chloride is about 1.2 times the stoechiometrically required

---

*)Patent applied for

amount. Use can be made of PVC for the production of HCl. Both through separate pyrolysis of PVC at 300°C, during which almost pure HCl is liberated, and through pyrolysis of PVC along with the metal to be treated, similar results are obtained. During the latter process also volatile hydrocarbons are released. Lead from soldered seams is converted to $PbCl_2$.

Similarly food cans which have been subjected to folding and subsequent strong compression can be detinned to a high extent without any trouble being experienced.

It may be expected that metals can also be extracted from other metal-containing waste materials and can be obtained separately by employing the process described. In the course of future experimental work attention will be paid to the processing of the organic fraction produced in the shredding of car wrecks. This fraction contains much polyvinylchloride and metals like Pb, Cd and Cu. Other waste materials that can be considered for treatment are paint waste, metal hydroxide sludge form the plating industry, fly ash, etc.

This work will be carried out in close cooperation with the Netherlands' industrial company ESMIL B.V..

THE OECD PROGRAMME ON HAZARDOUS WASTE MANAGEMENT

Pierre Lieben

Organization for Economic Cooperation
and Development
Paris, France

INTRODUCTION

For the last few years many countries, as well
as international organisations, have investigated the
technical problems associated with hazardous waste
treatment and disposal. A major study of this kind
was recently completed under the auspices of NATO/CCMS,
and is the focus of the present Symposium. The economic,
social and policy dimensions, however, have received
considerably less attention.

The importance of the economic and policy dimensions of hazardous waste management is now increasingly recognised by governments. The regulations that
have been passed and are gradually being implemented
affect the cost of production of the companies concerned, and may affect the production technology.
A country may not have the capacity to manage some of
the waste generated within its own boundaries, and
they may have to be transported across national
frontiers. The need to clean-up and rectify sites
where hazardous waste have been tipped imposes costs
either on public authorities or upon private developers.
Insurance for sites licensed to accept hazardous waste
may impose high costs on those companies or local
authorities operating these sites.

The amount of basic data required to examine
these various problems is extensive. It is necessary
to assemble some general information regarding the

regulatory environment in the different countries, as
well as the quality and nature of waste generated and
the expected increase in these flows in the future.
The current capacity of the hazardous waste treatment
industry, the methods of treatment and disposal and
their costs are also important parameters for any
assessment of the hazardous waste management problem.

ISSUES ASSOCIATED WITH HAZARDOUS WASTE MANAGEMENT

A number of economic and policy issues associated
with hazardous waste management were considered by
the OECD Waste Management Policy Group in 1980, during
the preliminary phase of development of its current
programme of work. They are briefly reviewed hereafter
and the programme which was undertaken in early 1981,
on the basis of this review, is then presented in
some detail.

Volume of hazardous waste generated and capacity of
the waste treatment industry

The volume of hazardous waste with which any
country has to deal, and the components of this waste
stream, are important variables in the management of
this category of waste. The introduction of a certain
type of treatment and disposal technology may depend
on a minimum throughput of certain categories of waste.
If not all generated in one country, some of this may
be imported. Not only the quantities generated on a
national basis should be considered, but flows between
regions should also be recorded and the regulations
governing their transport examined. In addition the
expected changes in the volume of waste to be treated
should be discussed. The volume of waste currently
generated will obviously represent an important input
into the decision process, but the volumes expected in
5 or 10 years time are also information of importance
to policy makers and planners.

Compliance costs for particular industries

The magnitude of the cost increases imposed on
particular industries is to be seriously considered.
There is concern in some countries that the control
of hazardous waste raises the costs born by certain
industries to the extent that the viability of these
industries will be seriously affected. For example,

the metal-finishing industry of several countries
consist of small firms which operate with obsolete
equipment in old premises. The extra capital and
operating costs imposed by more stringent hazardous
waste regulations may threaten the survival of some
of these firms.

There is also concern that small firms may be
absorbed by larger companies in the industry branch,
thus leading to a degree of monopolisation which may
be felt undesirable even if the net result of such a
concentration is a significant reduction in the total
volume of hazardous waste generated.

The nature and components of the cost increases
imposed on industry by more stringent regulations is
also to be investigated. In some countries there is
a particular concern about the increase in transport
costs caused by requirements to take certain waste
only to a limited number of approved treatment or
disposal facilities. In other cases hazardous waste
management may impose higher operating costs by requiring processes to be operated more efficiently. All
these costs are likely to affect the profitability of
the industry and, therefore, its competitiveness in
international trade.

Transfrontier transport of hazardous waste

There are several economic reasons for the transport of hazardous waste across regional or national
frontiers. These include the minimizing of transport
costs, the need to dispose of a particular waste
through a facility of significant scale (particularly
relevant for smaller countries agreeing to utilise a
single large facility rather than operating several
plants at low capacity and high cost), or the fact
that no facility is available in the country to treat
particular types of waste. In addition to the gathering of information on the types and quantities involved
in such "international trade", and the reasons for it,
a number of policy issues are raised which have to be
considered in detail. These relate in particular to
the procedures involved to permit the transport
between countries whose requirements for hazardous
waste management may be different, and the legal
provisions covering liability in case of an accident
during the transport or storage of waste outside its
country of origin.

Transporting hazardous waste across frontiers
also raises political and social problems. While the
general principle, in accordance with the Stockholm
Declaration on the Human Environment and the OECD
Recommendation on Transfrontier Pollution, should be
that hazardous waste is properly handled whether it
remains in the country of origin or is exported for
treatment and disposal, many people are particularly
concerned about what they conceive as the exporting of
an environmental problem from another country into
their own.

Siting of hazardous waste treatment facilities

Amongst the problems presented by the management
of hazardous waste, public reaction to the siting of
facilities is one of the most sensitive politically,
and one of the most divisive socially. Several countries have encountered particular difficulties in this
area. Ways of understanding and solving this problem
include the public enquiry and the environmental
impact assessment. The essence of these methods is
both to assess the environmental hazards that may arise
from hazardous waste treatment or disposal facilities,
and to allay the subjective concerns about the potential for damage to nearby populations.

This problem contains a number of sociological
and political aspects, and is very strongly dependent
on local conditions. A comparison of procedures used
in different countries, including their degree of
success in each application, would however add significantly to the understanding of siting policies and
to their improvement. For example the approach that
has been used in the past to evaluate the environmental
damage from noise or air pollution could be extended
to assess and correct the effects of proposals to site
hazardous waste facilities in a particular location on
house prices in that location.

Financial insurance and liability problems

There are several points in the hazardous waste
generation, storage and treatment/disposal cycle at
which problems of insurance and liability may arise.
However, the lack of actuarial experience on the likelihood of problems actually occuring presently limits the
extent to which both private and social needs for
appropriate insurance can be met through normal commercial channels.

The operating and "post-closure" insurance on licensed hazardous waste disposal sites and treatment facilities should be designed to ensure against malfunction of these facilities whilst they are in operation and, in the case of disposal sites, after they have been closed. Abandoned "problem" sites, like Love Canal in the USA or Lekkerkerk in the Netherlands, have recently thrown light and attracted considerable public attention on this particular issue.

The insurance against spills or accidents during transport may raise particular problems, and a great deal of popular concern, both because of the greater vulnerability of materials in transit and because, whilst being transported from one point to another, they may pass through areas of high population density. It is in such areas that spills of waste may cause the most significant problems either in the short-term, resulting from acute effects, or in the long-term, resulting from chronic exposure to toxic substances spread in the ambient environment.

The insurance for storage of hazardous waste should cover storage at the point of generation, whilst awaiting transport to a treatment facility, and storage whilst awaiting treatment or disposal at the appropriate facility. These are essentially short-term insurance problems. However, when long term storage has been chosen as a management alternative for certain categories of waste, like in the salt mines in Germany, the problems may be quite different. Very stringent requirements for the storage, or government-supported insurance schemes, may substitute in part for the absence of commercial insurance.

Administration and monitoring costs

Among the costs associated with hazardous waste management policies are those of monitoring and enforcement, whether at the national, regional or local level. The nature and extent of hazardous waste regulations in many countries suggest that these administrative costs may be significantly higher than in the case of other environmental management policies. Alternative policy approaches (voluntary versus regulatory, use of economic incentives, etc.) should be investigated in this respect. At a time when cost-effectiveness in the service provided by governments is being increasingly emphasised, the magnitude of enforcement costs as a proportion of the total direct costs of operating the regulations or guidelines appears particularly relevant.

THE CURRENT OECD PROGRAMME

Scope and objective

The programme of work that the 24 OECD Member
countries have decided to undertake on the management
of hazardous waste is directed towards the examination
of the economic, administrative and policy aspects of
the implementation of hazardous waste legislation, with
the ultimate objective of developing policy guidelines
for governments' consideration and proposing ways to
deal with problems of an international character. The
difficulty of finding appropriate sites for treatment
facilities, the remedial actions to take care of
abandoned sites where hazardous waste has been mismanaged in the past, and the contribution that economic
instruments can make in addition to regulatory measures
would also be considered.

Three priority areas were selected for the initial
phase of the work programme, undertaken in early 1981:

- the transport of hazardous waste across national
  frontiers, the quantities transported, the
  reasons for transport, and the practical and
  regulatory provisions necessary to ensure the
  safety of the operations;

- the costs to industry of complying with the
  regulations, and the costs to the regulating
  agency for enforcing these regulations;

- the financial provisions necessary to ensure
  the appropriate handling, storage and treatment
  of hazardous waste, including the monitoring
  of landfill sites for a sufficient period of
  time after closure, and the liability for damage
  caused by inappropriate management at any stage.

Method of implementation

The programme is to be carried out through the
Waste Management Policy Group, one of the working
groups of the OECD Environment Committee. Eighteen
Member countries currently participate in the regular
meetings (usually twice a year) of the Waste Management
Policy Group, and several international organisations
are also represented. The delegates are central government's representatives, responsible for waste management
policy in their respective countries.

The Environment Directorate of the OECD, with headquarters in Paris, is responsible for conducting the necessary studies, preparing the reports, and proposing conclusions and guidelines. This work is carried out with the help of expert consultants, under the guidance of the country delegate group and its parent Committee.

The findings are usually made available as OECD publications, once they have been finalized by the responsible committees, and in some cases the Council of the Organisation may adopt formal recommendations to Member countries.

Review of progress made

Abandoned sites. In November, 1980, during the preparatory phase of the current work programme, a technical seminar was held at OECD to address the problems raised by closed or abandoned "problem" sites where hazardous industrial waste has been mismanaged in the past, with a view to putting together and discussing the practical experience gained by Member countries in dealing with such problems. It was attended by some 35 experts from eleven countries. Three subgroups met in parallel to consider: (i) the location and identification of sites; (ii) the environmental and health impact assessment; and (iii) countermeasures and remedial action.

A comprehensive report of the findings of the seminar has been prepared and should be available shortly as an OECD publication. The main conclusions and recommendations from the experts will be considered by the Waste Management Policy Group at its next meeting, in November this year, for decisions concerning action to be taken, including the continuous exchange of information on hazardous waste "problem" sites through national delegates.

Survey of present legislation. A survey of the legislation presently in force in Member countries to control hazardous waste has been undertaken as a first, necessary step towards implementing the work programme as defined above. This survey will specify the regulations and guidelines with which those who generate, transport and dispose of hazardous waste are required to comply. It will include a comparative analysis of the dates of introduction of the various relevant provisions, the primary philosophy behind the policy

(for example, voluntary compliance, or clearly specified
mandatory provisions, or economic measures), and the
types of waste included, and will point to the similarities and differences among the various legal regimes.

A first draft of this survey was examined by the
Waste Management Policy Group in April, 1981. It is
presently being revised, corrected and supplemented
for further consideration by the Group next November.
It is expected that a final report could be available
in early 1982.

Preliminary examination of priority areas. The
collection of background information necessary to study
the three priority areas selected for the initial phase
of the work programme (transfrontier transport; compliance costs; liability and insurance problems) is well
under way. It indicates that hazardous waste management
policies in OECD Member countries are in an active
state of development, and that the time is ripe for
fruitful discussion of the implications at the international level.

In the field of transfrontier transport, the
information being assembled includes:

- the quantities (and possibly the categories) of
  waste imported, exported, and treated at sea;

- the reasons for exporting waste, and the destination of exports or origin of imports;

- the procedures followed and the documents
  required (import/export licences, notification
  procedures ...);

- the provisions for transit through the country.

Concerning compliance costs the study is considering the following data:

- long-distance transport costs;

- price lists for acceptance of waste into landfill
  and specialised facilities;

- "manifest" or "trip-ticket" systems (documents
  and explanations);

- government financial assistance for the establishment and operation of facilities;

- public enforcement costs (in terms of personnel
  involved) such as the operation of registration
  systems or the inspection and approval of sites.

On liability and insurance problems, the information
considered is mainly concerned with:

- the provision of third-party liability insurance
  in case of accident;

- the determination of liability for malpractice
  or for damages resulting from accidents;

- the provision of adequate funds to ensure proper
  operation of treatment facilities, decommission-
  ing and post-closure maintenance;

- the compensation of victims for damages arising
  from abandoned sites.

The study is expected to go on in 1982, with
reports completed by the end of that year.

NOTE: The views expressed in this article are those of
the author and not necessarily those of the Organisation
for Economic Co-operation and Development.

HAZARDOUS WASTE PROBLEM SITES

Robert D. Stephens

Department of Health Services
Berkeley, CA, United States

INTRODUCTION

Industrialized countries are faced with developing programs
to deal with closed or abandoned waste sites that contain a wide
variety of hazardous wastes either polluting ground and surface
waters or posing other unacceptable threats to human health and
the environment. These programs will typically involve identification of the substances at such sites and attempt to assess
their health and environmental hazard. Subsequently, there may
be an emergency response, or a long-term remedial action to
neutralize, contain, or remove the hazards. Methods to perform
these tasks are generally crude, very expensive, or ineffective.
Efforts are needed to identify or develop the most economical
methods for solving these problems.

Under auspices of the Organization for Economic Cooperation
and Development (OECD), a study was conducted in 1980 to address
the scientific and technical issues relating to problem hazardous
waste sites. Initially, this program involved extensive discussions with member countries on hazardous waste problems. The
discussions were held between OECD staff and with individuals or
groups engaged in research, in regulatory activities, in financial
administration, or in policy formulation. The purpose of the
discussion was fourfold:
- ° To assess the concern and related priority which member
  countries placed upon the problem hazardous waste sites.
- ° To identfy the important scientific and technical issues
  related to hazardous waste sites.
- ° To identify experts within member countries with experience and knowledge on the scientific and technical issues.
- ° To solicit participation in an OECD-sponsored seminar to
  deal with the identified issues.

Fifteen member countries were contacted relative to the
four points described above. Eleven member countries were visited
by OECD staff for purposes of direct discussions with national
experts. As a result of these discussions, a sense of priority
and urgency in dealing with hazardous waste sites became apparent.
Many countries had only just began to address this problem but
had already committed significant resources toward potential
solutions. A keen interest was expressed by many member countries
in participation in a seminar of experts representing the OECD
countries. The purpose of such a seminar would be to focus on
specific important scientific and technical issues related to
problem hazardous waste sites. The seminar would produce a
report which would:
   ° Discuss the sum of the scientific and technical knowledge
     on the specific laws.
   ° Identify the major gaps in the scientific and technical
     knowledge important to the solution of adverse impacts
     from problem hazardous waste sites.
   ° Make recommendations as to where productive research and
     technical development might be carried out, particularly
     by those areas amenable to multinational efforts.

An important aspect to this project was the focusing of
discussions on the salient technical issues. Much of this process
was accomplished by OECD staff prior to the conduct of the actual
seminar through visits and direct discussions within OECD member
countries. As a result of these discussions, three issues emerged
as key components to the overall problem of hazardous waste
sites. The three components were that of site discovery and
location, site assessment, and remedial measures. These issues
would then provide the structure around which the seminar could
be organized into three subgroups each taking as a focus one
of the three subject areas. These three topics, which are discussed in detail below, of site discovery, analysis, impact
assessment and corrective measures, were then established as
subject areas for discussion by three subgroups of the experts
seminar. Further refinement of each of these subject areas
would be accomplished during the seminar. It was recognized
that a certain amount of overlap existed between the three subject
areas. Site discovery requires some degree of environmental
assessment. Site impact assessment requires consideration of
location and potential remedial actions. Remedial measures must
consider questions of environmental chemistry and health impacts.
The overlap of the three subgroup subject areas was seen as a
benefit in the preparation of a cohesive final report composed
of the work of the three separate subgroups.
    The seminar of experts was convened in November 1980. The
participants were assigned to one of the three workgroups based
upon expertise and experience. Each subgroup had a chairperson
who prepared a working paper to organize and direct the discussions of the group. These working papers, which were to provide
structure to the final report of each subgroup, can be found in

the appendix of this report. The chairpersons of each subgroup
and their affiliations were as follows:

Subgroup 1: LOCATION AND IDENTIFICATION OF PROBLEM SITES
             Chairman: M. Alain Perroy, Agence Nationale
             pour la Recuperation et l'Elinination des
             Dechets, France.

Subgroup 2: ENVIRONMENTAL AND HEALTH IMPACT ASSESSMENT
             Chairman: Mr. Howard Hatayama, California
             Department of Health Services, U.S.A.

Subgroup 3: COUNTERMEASURES AND REMEDIAL ACTION
             Chairman: Dr. John Bromley, Hazardous Materials
             Service, Harwell Laboratory, United Kingdom.

SUMMARY OF FINDINGS OF THE SEMINAR

Overview of Subgroups 1, 2, and 3

        The seminar was initiated by a sharing, in each subgroup,
of the direct or indirect experience of each member of specific
problems, solutions, and approaches to problem hazardous waste
sites in their respective countries. This sharing gave evidence
of a number of common general aspects worthy of note which are
not presented in the specific recommendation of each subgroup.
     ° The occurrence of or the awareness of problem waste
        sites is relatively new in most countries, and scientific
        and government institutions were experiencing considerable
        difficulty in meeting the challenge.
     ° The location and discovery of many problem waste sites
        without the institutional capability to deal with the
        resultant problems has caused significant public anxiety.
     ° Actual environmental and public health impacts resulting
        from problem waste sites is, in general, either very
        poorly understood or misunderstood. The scientific founda-
        tion on which particular health impact assessments are based
        is weak. As a result, development of such health impact
        measurements will probably be some time in development.
     ° Environmental and health impacts are complex and far-
        reaching, remedial and countermeasures are uncertain in
        the long term and often exceedingly expensive. The most
        practical and effective solution lies in avoidance of
        such problems before they occur. Avoidance of problem
        waste sites comes from minimizing the placement of "problem
        wastes" in the ground and the proper engineering and
        control of those absolutely necessary land disposal sites.
     ° Considerable complementary scientific and technical
        knowledge exists within the participating OECD member
        countries. Information of immediate and future use in
        dealing with specific problems relating to problem waste

sites was transmitted and received by participants. A
keen interest was expressed in the continuance of such
multilateral exchanges. Particular interest was expressed
in comprehensive discussions of expecific issues.

FINDINGS OF THE EXPERT SEMINAR

Subgroup 1--Problem Site Identification and Preliminary Risk Assessment

Overview. The majority of participants of subgroup 1 supported the concept and importance of a rational well-planned inventory
of problem hazardous waste sites. With the exception of programs
described by the French and the German representatives, however,
little actual experience in such survey programs existed. Several
group members described site survey programs which contained no
environmental or health impact criteria. Such programs had
identified hundreds or even thousands of potential problem sites.
General agreement was reached within the subgroup that such
non-selective surveys were of limited value. A key aspect to a
meaningful survey was therefore seen as adequate measures or
estimations of potential environmental and health impacts.
Such measures would allow for a realistic ranking of located
waste sites for further assessment. Considerable practical
difficulty, however, is encountered in establishing such ranking,
for impact assessments which are readily conducted on large
numbers of potential problem sites are generally unreliable, and
carefully conducted assessments are too expensive and time consuming to be of practical value.

The success of problem site inventory programs was seen as
dependent upon or greatly influenced by the degree of cooperation
between the private waste-producing land-disposing sector and
the public sector. Government institutions had great difficulty
in obtaining sufficient accurate information about waste sites,
their character, and location without the cooperation of waste
producers.

Specific Findings. The location of problem sites is a
consequence of the development of hazardous waste management
regulations and standards. The development of criteria defining
hazardous or special waste, the environmental and public health
impact such waste causes, and the mitigation of these impacts,
brings to public and governmental attention the problems caused
by waste generated and disposed in previous times.

Considerable information exists within the public and private
sector regarding waste activities of the past. There is, however,
a great need to conduct systematic inventories and data collection
on sites; otherwise data sets become too large and cumbersome to
handle.

Guidelines for initial risk assessment, the factors to be considered, and the methods by which data is to be collected and weighted are needed. Preliminary risk assessment often requires environmental sampling. Policy or guidelines governing such sampling is needed in light of the very large numbers of problem sites initial inventory programs identify.

Participation by and effective communication with affected community groups is needed to prevent undue anxiety or adverse public reaction which makes rational solutions to problem sites more difficult.

Most countries lack proper legislation and government resources to deal with problem sites. Although the responsibility for redressing these problems lies primarily with the private sector, government has an important leadership and regulatory role to play.

A continuing exchange of information between countries on technical, legal, and philosophical issues will contribute significantly to the solution of problem sites.

Subgroup 2--Environmental and Health Impacts from Problem Sites

Overview. The collective view from members of subgroup 2 was that quantitiative assesment of environment and health impacts from waste sites is an exceedingly difficult task. In addition much of the basic science which would support such assessments has not been done. This basic science relates to the effect of long-term exposure to subacute levels of toxins, the basic role of chemical carcinogens, and the synergistic and antagonistic interactions between different toxins. The complexity and sheer numbers of physical, chemical, and biological parameters potentially measurable at a waste site is overwhelming. The practicality and usefulness of such large amounts of data is questionable. There is, however, an undeniable need for good quality, reliable physical, chemical, and biological data. Effort is needed, therefore, to develop an effective assessment system which iterates the most useful and accessible environmental and health parameters.

Ultimately, impact assessments become or are part of risk assessment. Risk assessment requires criteria values. Many difficult philosophies and approaches to such criteria were expressed by different national representatives. The establishment of standard multinational criteria may be quite difficult. However, if an effective system for relating meaurable contamination to environmental and health effects is available, then substantive discussions leading to such criteria could begin.

Specific Findings. A considerable knowledge base exists on geological and environmental contamination assessment. The use

of all of these available methods makes problem site investigation
exceedingly expensive and protracted. Guidelines need to be
developed as to how specifc information might be ranked as to
importance in impact assessment. Such ranking must of course be
a function of each site under investigation. Priority ranking
of environmental data is complicated due to the complex nature
of most waste-related contamination. Behavior and impact of
single substances is quite well understood, whereas that of
mixtures of substances is not. The use of surrogate parameters
in environmental measurements is useful but risky. Quality
assurance programs are mandated.

The use of biological measurements, both acute and long-term
subacute, should prove very useful for impact assessment, particularly in coordination with physical and chemical measurements.

Insuffucient means exist to assess exposure to, and effects
of, low-level, chemical toxins. Methods need to be developed
which are rapid, reliable, and noninvasive. Such methods would
be applicable to large and small public groups potentially im-
pacted. The existence of such routine assessment methods would
allow for a health monitoring program.

Much established epidemiological methodology is designed
for large study groups; it is not applicable or accurate for
small groups. Some applicable methodology and expertise exists
in the occupational health community which might be adapted to
localized environmental health assesment.

Much of the health impact from problem sites may be psychological and not physiological. Such adverse psychological effects
result from technical misunderstandings, high-level media coverage,
or other factors not related to toxic contaminants.

Subgroup 3--Short- and Long-term Remedial Measures

Overview. Problem sites in this group were identified as a
source of high-level contamination. The group consensus was
that remedial efforts at problem waste sites were directed toward
one of only two options. These options were site isolation or
encapsulation with subsequent water management, or contamination
removal with subsequent redeposit or treatment. The choice
between these two options must be based upon economics, risk
assessment, and physical feasibility. Assessment of risk as a
result of excavation, transportation, and treatment of toxic
chemicals is often feasible, based upon industrial experience in
handling such materials. Risk assessment relating to long-term
land storage is more difficult and often less precise.

Discussion of a variety of remedial measures at problem
waste sites as decribed by the group's participants highlighted

the extreme expense and difficulty by such actions. Examples of
both isolation and excavation were presented. In the overwhelming
majority of the situations presented, the problem site could
have been easily and effectively prevented by proper waste management at time of generation and disposal. Cost of treatment and
proper disposal of problem waste at the time of generation would
appear to be only a small fraction of the cost of redressing
problems resulting from improper disposal. This fact created an
imperative to ensure that problem sites which require remedial
action are not being created currently by the use of improper
disposal methods.

    <u>Specific Findings</u>. Technology for site encapsulation exists;
however, applicability to many waste types and and long-term
stability need to be established.

    Cost effectiveness of site encapsulation needs to be addressed,
as the cost of such methods may in the long term exceed the cost of
complex removal.

    Encapsulation, in certain cases, is mandated due to contami-
nation potential and infeasibility of excavation. Technology
for ensuring long-term stability of encapsulated sites is not
completely adequate. Areas in need of further development are:
leakage control of surface sealing, stability of materials against
encapsulated substances, improved methods to increase impermeabil-
ity of native soils or decrease leachability of contaminant
substances, and improved long-term, low-cost monitoring methods.

    Technology is generally available for management of surface
waters; however, such methods are often very expensive. Management
of groundwater is less well understood, and currently available
methods are not completely satisfactory. Decontamination of
groundwater with currently available technology is very difficult.
Behavior of contaminants in complex hydrogeological systems
(i.e., multi-layer aquifers) is poorly understood. Research in
this area is badly needed.

    Excavation of waste is commonly at high cost and high risk.
Transport of excavated materials to remote locations for treatment
and/or disposal may involve long distances and international
frontiers. Lack of sufficient disposal capacity or expense of
treatment usually limits excavation to relatively small areas of
contamination.

    Methods of risk assessment for comparing the different
technologies for problem site remedial action need development.
Such risk assessment should share a common value base within the
OECD community, since environmental impacts from problem sites,
or from remedial measures at problem sites, often affect neighbor
countries.

Considerable and varied experience exists within the OECD community about remedial technologies. Continued exchange of this technology, its cost effectiveness, and its long-term success would be valuable to member countries.

RECOMMENDATIONS

Recognizing that this expert seminar on Problem Hazardous Wastes Sites has established that problem sites exist in virtually all industrialized countries;

Recognizing that scientific and technical procedures, standardized criteria and guidelines, and centralized data bases do not generally exist within most countries at this time; and

Recognizing that resources within countries are generally limited for the identification, investigation, and remedy of problem waste sites, as well as for the research and operating programs required to improve procedures and to standardize criteria and data gathering;

The Waste Management Policy Group, in reviewing the technical findings of the expert seminar, recommends that member countries:

(1) Examine their legislation and government resources to determine if they are adequate to deal with problem sites or at least to provide a leadership and oversight role.

(2) Participate in a multilateral program organized under OECD or another appropriate international pilot-country mechanism which would:
   -- develop guidelines for minimum data collection for the identification of problem waste sites.
   -- develop guidelines for the initial and subsequent investigations and risk assessments at the site.
   -- exchange information on experiences in site identification, investigation, and assessment.
   -- review ongoing research programs which are developing site sampling and analytical procedures, biological measurement methods, exposure assessment methods, surrogate parameters, epidemiological methodology, and remedial engineering methods such as site encapsulation, waste removal, or waste destruction.
   -- determine where bilateral or multilateral cooperation could enhance the ongoing research programs.
   -- compile a data base on cost and effectiveness of remedial measures which have been applied.

HAZARDOUS WASTE INCINERATION

Christian Nels

Federal Environmental Agency
Berlin (West), Federal Republic of Germany

1. INTRODUCTION

The problems of hazardous wastes have increased in recent decades in line with the increase in the demands of consumers and with the consequent growth in industry. At the same time, however, it has become accepted that efforts to manage these wastes should include elimination or reduction of these wastes or recycling them into production processes. Nevertheless, whatever recycling endeavours may be made, a certain proportion of hazardous wastes will always remain for disposal.

In order to sustain our ecosystem it is necessary to minimize the negative influence of hazardous waste disposal by transforming waste substances within a limited amount of time into a form where they can be discarded without any damage to the environment. In addition to physical and chemical treatment, incineration appears to be a viable option.

The value of hazardous waste incineration must be considered in the light of an increased awareness of environmental protection (manifested in corresponding environmental protection legislation) and the need to use energy and raw materials economically. Thus, incineration with heat recovery should be the target for wastes which could not be otherwise used but could be safely incinerated.

2. ASSESSMENT OF HAZARDOUS WASTE INCINERATION

The decision to choose incineration for hazardous waste treatment rests on various convictions:

- Certain toxic or harmful contaminants can be converted into harmless compounds by oxidation in the incineration process. Thus, the
  long-term risks associated with landfill disposal can be avoided.
- Incineration of hazardous waste is an excellent method of volume reduction. This is an increasingly significant factor at a time when the
  availability of landfill sites is decreasing.
- In any selection of a technology for waste treatment, energy use is
  a major factor especially with the upward trend of energy prices. Organic wastes with sufficient calorific value should therefore preferably be disposed of by incineration with heat recovery to utilize the
  heat content of the waste. It will be necessary to intensify research
  work in order to investigate the technology and economics of heat
  recovery from hazardous waste incineration.

Having stated the desirable characteristics of incineration, it should
be kept in mind that

- Emissions from incineration have to be subject to control to eliminate
  particulates, noxious gases and other hazardous components.
- Effluents from flue gas purification (e. g.,saline water, slurries)
  may arise for disposal.
- The residues such as slag and fly ash themselves may be classified
  as hazardous wastes and in some cases require special treatment
  (e. g.,fixation).
- By-products may be generated, which need further investigation because of their high toxicity.

These factors often make incineration a more expensive disposal
route, especially if it is compared with landfill. The rather limited use
of incineration to date indicates that it has been seen predominantly from
a short-term economic view point: wastes currently are incinerated only
if they have a sufficiently high calorific value for incineration. The question remains whether incineration should not be increasingly assessed in
the light of ecological criteria i. e. seen as a technical means of degrading
hazardous wastes or their chemical compounds into less hazardous or non-
hazardous compounds within a controlled period of time by the effect
of high temperatures, and whether purely economic aspects should not
be considered of lesser importance.

3. TYPES AND VOLUMES OF INCINERATED WASTES

In order to assess the feasibility of incineration for a given hazardous waste it is crucial to know about its characteristics the physical,
and thermo-dynamic properties. In general it can be stated that wastes
containing carbon, hydrogen and/or oxygen are good candidates for incineration with appropriate incineration technologies. Wastes containing
higher chlorine percentages (exceeding 30 % by weight) as well as
other halides, phosphorus, sulfur or nitrogen are less suited because they
require more sophisticated incineration technologies. The incineration
of wastes containing a high portion of inorganic compounds and/or metals on the contrary is not suitable.

are: $_{1)}$Other factors to be considered in evaluating waste for incineration

- Moisture content
- Potential pollutants present in incinerator effluents
- Inert content
- Heating value and auxiliary fuel requirements
- Potential health and environmental effects
- Physical form
- Corrosiveness
- Quality
- Known carcinogenic content
- PCB content.

Experience shows that the main types of wastes delivered to the incineration plants may be listed up in the order of their frequency as follows:
- solvent wastes and sludges
- waste mineral oils
- varnish and paint wastes and sludges
- plastics, rubber and latex wastes
- oil emulsions and oil mixtures
- phenolic wastes
- mineral oil sludges
- glue, putty and non-congealed resin wastes
- grease and wax wastes
- pesticide wastes
- organic wastes containing halogen, sulphur or phosphorous compounds
- refinery wastes incl. acid tar and spent clay
- drilling and grinding emulsions

The specifications concerning types of wastes not accepted can be generalized only with regard to explosive substances and highly radioactive substances.

Other restrictions must be seen in the context of the individual wastes and individual plants.

In order to get an idea of the volume of incinerated wastes the following table may be presented as an example. The information given should be critically assessed, because the assumptions vary from country to country as can be seen from the numerous footnotes.

---

1) An extensive overview over various kinds of hazardous wastes and their suitability to different types of incinerators is given in chapter 3 of the "Engineering Handbook for Hazardous Waste Incineration (Draft)", U.S. EPA.

Table 1: Survey of the occurence of hazardous wastes in individual countries

Country	Total volume of waste (t/a)[1]	Portion incinerated (t/a)	(%)
France	2.000.000	400.000[2]	20
Germany	3.500.000	540.000[3]	16
Great Britain	3.000.000	150-200.000[4]	5-6,6
Netherlands[5]	420.000	210.000	50
Norway	200.000	50.000[6]	25
United States	35.000.000	1.960.000	5,6[7]

1)   including waste oils

2)   not including on-site incineration

3)   not including appr. 320.000 t/a which are co-incinerated in household waste incineration plants

4)   not including in-house (on-site) incineration of about 600.000 t/a in 73 in-house plants of which substantially more than 50 % is accounted for by the oil refineries and the largest chemical manufacturers

5)   including waste oil and substitute fuel, including incineration on sea, excluding biological sludges, excluding on-site treatment

6)   including 50 % of waste oil used as fuel substitute

7)   another 9,7 % go to uncontrolled incineration (all figures being estimated)

4. LEGAL CONDITIONS FOR INCINERATION

The legal conditions for incineration vary from country to country. Whereas in some countries incineration plants for hazardous wastes are approved on a case-by-case basis, in other countries there are uniform countrywide regulations. These standards may go as far as to cover requirements for site location and design, operating methods, emission limits, control of residues, contingency plans, personnel training, financial responsibility, record keeping, reporting, monitoring, inspection and others. For example, in the Federal Republic of Germany, the construction and operation of hazardous waste incineration plants is approved under a plan approval procedure. The Federal Government has issued general administrative regulations, regarding particular

-    immission limits which, to achieve the prevention of harmful effects, may not be exceeded,

- emission limits which, in the light of the latest state of technology, need not to be exceeded,
- the procedure for the establishment of emissions and immissions.

With special regard to environmental protection Table 2 gives details of the emission limits set for some individual plants in the Federal Republic of Germany.

Table 2: Examples of emission limits and volumes (mg/m^3)

No. of plant	Year of commissioning	CL^- Volume	Limit	F^- Volume	Limit	SO_2 Volume	Limit
1	1981	< 50	100	< 2,5	2,5	< 200	200
2	1967	< 250	1.000	< 10	15	< 350	2.500
3	1976	< 50	100	< 3	10	< 300	1.250
4	1960-1978[2]	< 90	100	< 0	-	< 300	-
5	1977[2]	< 100	100	< 5	5	< 2.100[1]	2.100[1]
		NO_x		CO		Dust	
1		< 400	400	<100	100	< 75	75
2	See	<400-800	1.000	< 30	100	80-120	150
3	above	<400	1.000	<100	100	< 20	100
4		< 200	-	< 80	100	< 100	100
5		-	-	-	100	< 100	100

1) together with a combined oil-fired steam boiler

2) additional limits for C_{org} 50 mg/m^3

Comparison of emission limits with the year of commissioning shows almost without exception a progressive tightening of standards followed by a corresponding reduction in the amount of harmful constituents emitted, which verifies the high efficiency of environmental protection devices.

5. TECHNICAL FEATURES OF INCINERATION

As with the incineration of domestic refuse, the incineration of hazardous waste has been used as a method of treatment for many years. There has as yet been no systematic development regarding the technical features of plants, the layout, or even design bases which is comparable to that seen in the field of domestic refuse incineration. This is chiefly due to the wide variety of hazardous waste substances which, in some cases, exhibit divergent incineration properties and necessitate the use of installations which specifically match these. There is thus more than one solution to the problem of incinerating hazardous wastes.

5.1 General Overview

The following section is not aimed at giving a complete description of hazardous waste incinerators but at discussing some details which seem to be of major importance. The rotary kiln (Fig. 1) has established itself as an universal incineration system. It is suitable for the simultaneous incineration of solid, liquid and semi-solid wastes of a wide range of calorific values. It is, however, not the best solution for the sole incineration of liquid or gaseous wastes. For this purpose liquid injection (Fig. 2) is more suitable. Although the simple firebox is also suitable for all types of wastes, its throughput capacity is considerably reduced in the case of hazardous waste with a high solid residue content (slag/ashes) since automatic slag/ash removal presents complications. Grate incinerators for solid wastes are mainly used to burn household refuse, and are not often applied to hazardous waste disposal unless for co-incineration. Multiple hearth incinerators are mainly used to burn Sludges, tars and hydrocarbons. They may become of major interest for

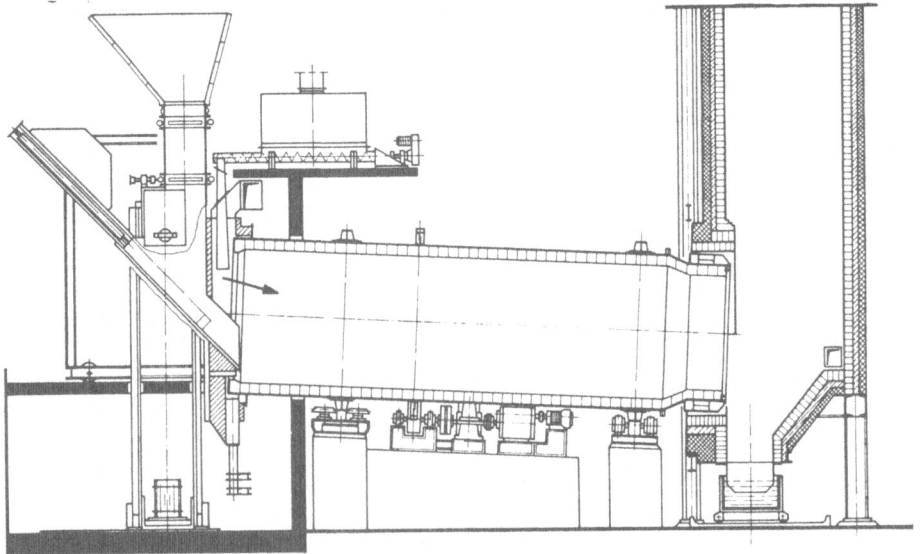

Figure 1:   Rotary Kiln

1	Burners	5	Waste heat boiler
2	Combustion air supply	6	Steam drum
3	Combustion chamber	7	Superheater
4	Smelt dissolving tank	8	Ash removal

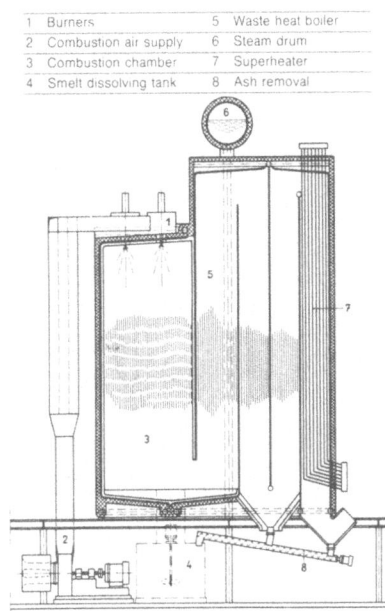

Figure 2: Liquid Injection

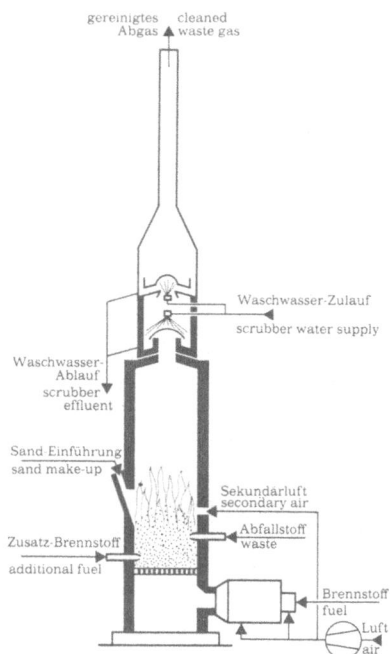

Figure 3: Fluidized Bed Incinerator

the combined incineration of sewage sludge and some types of hazardous wastes (e. g.,pesticides). Fluidized bed incinerators (Fig. 3) are very versatile, being used for the disposal of solid, liquid, and gaseous combustible wastes. The most common applications are in the petroleum and paper industries, and in sewage sludge disposal.

5.2 Description of a selected Facility

Figure 4 shows the process scheme of an incineration plant with flue gas scrubbers for the disposal of chemical wastes installed at the Bayer facilities in Leverkusen /Germany. This plant has been described in detail by Fabian, Reher and Schön (1979). In operation since November 1977, it burns about 25.000 t/a of industrial wastes, simultaneously producing nearly 140.000 t of steam. The plant consists primarily of the solid waste bunkers (1), the kiln charging system (2), the rotary kiln (3), liquid waste burners (4), the afterburner chamber with emergency stack (5), the heat recovery boiler (6), the electrostatic precipitators (7), and the flue gas scrubber (8), which comprises an injection cooler (quencher), two rotary scrubbers, and a jet scrubber. The scrubbing water of this jet scrubber is oxidised in (9). The cleaned flue gases are reheated in two heat exchangers (10) and discharged by a suction blower (10) via a 100 m high chimney (11) into the atmosphere. Around the chimney several measuring instruments are installed to check and continuously record the type and amount of emissions. The scrubber water is sent to the company's own waste water treatment plant. The solids are landfilled separately.

The system requires an operator staff of 35 persons including the instrumentation and maintenance personnel. The aquisition cost was DM 27 million. The operating cost is DM 10 million per annum. The average incineration costs are DM 400 per ton of waste, ranging from DM 165/t for liquid non-chlorinated waste to DM 885/t for waste in small containers of highly variable content (1979 prices).

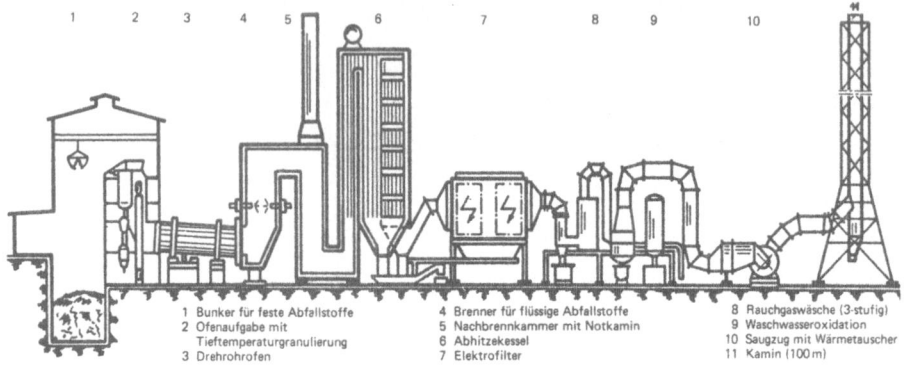

1 Bunker für feste Abfallstoffe
2 Ofenaufgabe mit
   Tieftemperaturgranulierung
3 Drehrohrofen

4 Brenner für flüssige Abfallstoffe
5 Nachbrennkammer mit Notkamin
6 Abhitzekessel
7 Elektrofilter

8 Rauchgaswäsche (3-stufig)
9 Waschwasseroxidation
10 Saugzug mit Wärmetauscher
11 Kamin (100 m)

Figure 4: Bayer Incineration Plant

5.3 New Developments

An incineration facility consists of three major areas: the waste acceptance and charging system, the incinerator system including the reactor, afterburner chamber (if necessary) and heat recovery boiler and finally the flue gas scrubbing system. This section cannot cover all new developments, but one example for every area is given, which is of some interest from the author's point of view.

Corresponding to the description of the Bayer facility, it should be mentioned that in July 1980 a severe explosion occurred in the incineration facility, which caused one death, several injuries and destroyed major parts of the bunker building and the adjacent incinerator. The reason for this accident was mainly a chemical after-reaction of liquid wastes stored in container. The development of heat combined with this reaction caused the release of burnable gases which were ignited by the waste burner.

As a result the operator decided not to restore the bunker building in its original pattern, which included four chambers having a capacity each of 200 m^3 for neutral, basic, acid and special waste types, but to go over to a container feeding system exclusively and to limit the volume of wastes, of which an after-reaction could not be totaly excluded, to 5 liters per container.

A type of reactor which so far has not gained major importance
for the incineration of hazardous wastes is the shaft furnace. The 'Jülich Incineration Process' may stand as an example for new developments
in this area. Whether or not it will come into common use in the future
is not obvious at the present time.

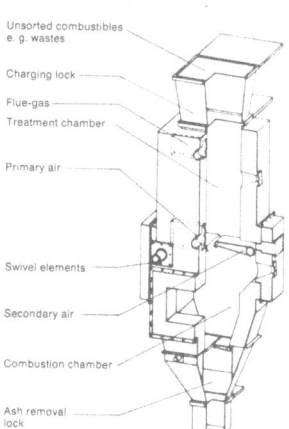

Figure 5: The Jülich Incineration Process

The 'Jülich Incineration Process' is a continuously operating two-
stage process. It permits the combined incineration of wastes with heterogeneous characteristics. Sorting-out or pretreatment is only required in
case of bulky waste. In the first process stage, the wastes undergo thermal
and mechanical treatment, producing pyrolysis gas and coke. Due to admis-
sion of secondary air these products are burnt out in the second process
stage. The maximum temperature in the treatment chamber ranges from
600 to 800 °C. In the combustion chamber it ranges from 800 to 1.000 °C.
Completely burnt out flue gas and unmolten ash is produced. Support firing
is not required if the average calorific value of the wastes in the treatment chamber is higher than 4.200 kJ/kg.

A considerable amount of research and development is currently
done in the field of flue gas cleaning. Whereas electrostatic precipitators
for dust removal and wet scrubbing system for the absorption of noxious
gases have proved reliable and are widely used, dry removal processes
are on the margin of achieving the necessary removal efficiencies at acceptable costs. In this connection a rather interesting method should be
mentioned, a so-called wet scrubbing system with a lined up dry reactor
(Figure 6). In a two-stage scrubbing system, acid gases such as HCl,

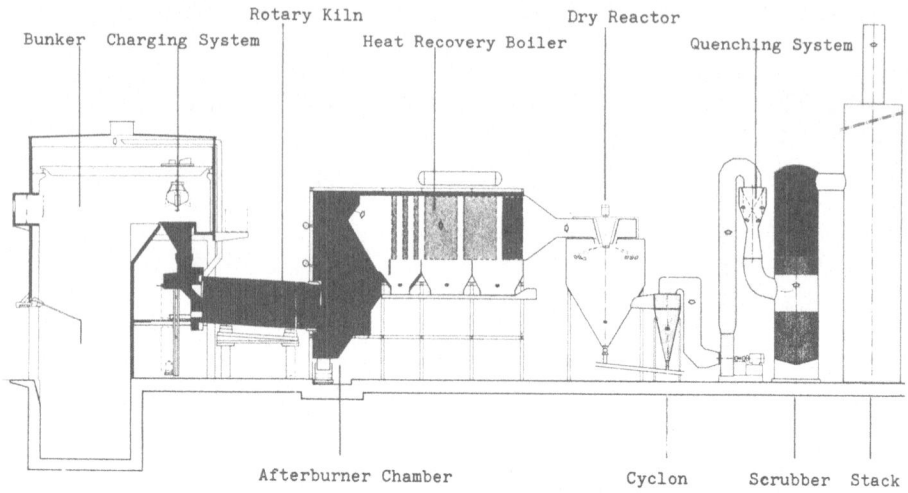

Figure 6: HIM - Hazardous Waste Incinerator in Biebesheim

HF and SO$_2$ are removed and fine dusts and aerosols are separated. The effluent produced in the scrubbing system is injected into a dry reactor located between the heat recovery boiler/electrostatic filter, on the one side, and the gas scrubber on the other. In this dry reactor the water evaporates due to the heat of the flue gas. Coarse solids are removed in a dry state in the reactor itself, and final solids in two cyclones connected up-stream of the reactor. Residual dusts are separated out in the subsequent scrubbers and fed back into the dry reactor with the effluent. It is hoped that this circulation will lead to an agglomeration of the solid harmful constituents into larger particles.

6. ENVIRONMENTAL ASPECTS OF HAZARDOUS WASTE INCINERATION

The chief environmental problems linked with the incineration of hazardous waste are as follows:
- air pollution
- pollution of surface waters
- pollution of the soil

The pollution can be considerably reduced by suitable equipment and corresponding measures.

6.1 Air Pollution

The main problems of air pollution are caused by the dusts and noxious gases which are emitted. The main substances which are toxic to man are:

- dusts, consisting primarily of silicate compounds and oxides, and containing heavy metals, generally in compound form (e. g.,Pb, Cd, Zn)
  or more rarely vapours (Hg),
- toxic gases, including oxides (SO_2. NO_x, CO) and acids (HCl,
  HF),
- organic micropollutants.

The composition and quantities of substances emitted to atmosphere
depend on several factors:
- the composition of the wastes treated,
- the anti-pollution equipment used and its efficiency,
- the incineration system used,
- the combustion temperature,
- the residence time of the gas burnt in the combustion chamber.

The removal of dust is possible with cyclones, bag filters, electrostatic cleaning systems, and wet scrubbers. Cyclones do not achieve a
sufficient removal efficiency, particularly with fine particles. Bag filters
demand extensive cooling of the flue gases and are very expensive. Although
wet scrubbers afford the possibility of simultaneously eliminating the noxious
gases, disadvantages are that these units require large throughputs of
water and considerable reductions in pressure occur. Moreover, the particulate matter must then be separated from the scrubber effluent in a further
stage in the process and the resulting sludges dewatered. Electrostatic
filters are in common use on account of the high removal efficiency required. Efficiencies from 98 % to over 99 % are being achieved.

In newly built plants in which the elimination of toxic gases becomes
necessary, the separation and dry removal of dust (e. g.,by means of
an electrostatic filter) and wet removal of noxious gases will become
more common. Various types of scrubber have proved reliable for the absorption of the noxious gases HCl and HF from the flue gases. All of these
make use of the solubility in water of the gases to be removed. If, in
addition, the removal of SO_2 is necessary, this requires the use of alkaline absorption agents on account of its lower solubility in water.

Table 3: Wet scrubbing removal efficiency

Plant	noxious gases	crude gas (mg/m^3)	clean gas (mg/m^3)	efficiency
A	HCl	7.450	30	> 99
	SO_2	6.255	600	> 90
	dust	8.000	30	> 99
B	HCl	2.400		≥ 97
	SO_2	4000-10000		~ 80

Table 3 gives the removal efficiency of wet scrubbing systems in two different hazardous waste incinerators in the Federal Republic of Germany.

Dry removal processes, which are based on the absorption of the toxic gases by means of suitable absorption agents (e. g.,Lime CaO) with subsequent dust removal, are the subject of current research and development, since from the point of water management the introduction of these processes would confer substantial advantages.

In the past analysis was mainly carried out for the more significant flue gas components such as mentioned in table 3. However, some other components have recently become the basis for concern from the incineration of hazardous wastes. In this context it may be difficult to remove volatile metals, such as lead, cadmium, zinc, mercury, arsenic and antimony from the effluent gases. As the gas cools, metal or metal oxide fumes may form, characterized by their very small particle size which is typically less than 0,1 micron. An available efficient gas cleaning system for such fumes is the bag filter although this has many disadvantages for use on an incinerator. Materials containing substantial concentrations of these metals are thus best not incinerated until better gas cleaning devices are developed.

Finally an additional environmental burden which can result from the incineration of hazardous wastes should be mentioned. There are several groups of thermally stable chlorinated organic compounds including the polychlorinated biphenyls (PCBs), polychlorinated dibenzodioxins (PCDDs), polychlorinated dibenzofurans (PCDFs), and hexachlorbenzene, which may originate from many combustion processes. In general they are not normally present in the fuel or waste, but are formed during the combustion process from aromatic and possibly aliphatic chlorinated organic compounds. Because of the very low concentrations in which they occur (0,1 ppm), the identification of these pollutants is very difficult. However, investigations carried out in The Netherlands, Switzerland and the US have confirmed the existence of some of the above mentioned compounds in the flue gases from incinerators.

There exist numerous isomers of the above mentioned compounds with different toxicities. Only a small proportion of the chlorinated pollutants emitted from waste incinerators are isomers with a high toxicity such as the 'Seveso' poison. However, the test results available are insufficient conclusive as to whether the nature and quantity of chlorinated organic compounds in incinerator emissions represent a health hazard. Detailed investigations and extensive monitoring are necessary to obtain reliable, representative results on which the danger of these emissions can be assessed objectively, and any programme aimed at reducing the concentrations can be based.

6.2 Pollution of Surface Waters

If it is discharged untreated into the receiving stream, the effluent produced by the wet scrubbing of the flue gases may cause damage because of its acid nature. The necessary neutralization process at any rate generates soluble salts such as $CaCl_2$, NaCl and NaF. Owing to the increasing burden of salts and heavy metals in rivers it is to be expected that in the future desalination of scrubbing water and precipitation of heavy metals will be increasingly required as well as neutralization and separation of solids.

6.3 Soil Pollution

There is a risk of soil pollution when solid residues fron incineration are landfilled. These are
- slag from the reactor and, where present, the afterburner;
- dusts from cyclones, electrostatic filters and other filters;
- scrap metal.

It should be kept in mind that bottom slag from hazardous waste incinerators, mainly rotary kiln reactors, differs considerably from bottom ash from household waste incinerators. Because of the generally higher process temperature (∿1.000°C versus ∿800 °C) and the more homogeneous nature of the input, the residues appear to be more slaggy or sintered. Elution of pollutants is therfore more likely to be limited. Fly ash from hazardous waste incinerators usually has the following properties:
- a high concentration of volatile toxic heavy metals,
- a high content of watersoluble compounds such as sulfates and chlorides,
- a low pH (acid) of the suspension.

It has been concluded that fly ash no longer should be disposed of together with slags and that high water solubility of some compounds must be reduced through suitable treatment such as compaction. Separate landfilling of fly ash and bottom slag is applied in the case of nearly all hazardous waste incinerators. The fly ash then is either stored in special containers, which go to a chemical landfill, if no treatment takes place, or it is disposed of after neutralization and fixation with lime in a chemical landfill.

7. COSTS OF INCINERATION

Table 4 gives relevant data for six modern German hazardous waste incinerators as compiled by the working group 'Flue gas Scrubbing' of the association of the German Chemical Industry (all prices in Deutsch Marks). As can be seen from this table, investment costs as well as operating costs are significantly higher than those of domestic waste incineration plants. This is related to the fact that the numerous ancillary installations for fuel preparation and pretreatment, flue gas cleaning, etc., necessitate a considerably greater outlay on structural work than in the case of domestic waste incineration plants. In summary it could be said that with hazardous waste incinerators about 10 - 17 % of the investment costs originate from flue gas scrubbing devices and that their operating costs amount to about 20 - 40 % of the total operating costs.

The information given in table 4 is indicative rather than conclusive and should be used with caution. While translating the figures into other

Table 4: Cost Comparison

Plant		A	B	C	D	E	F
Thermal Load	GJ/h	100	80	115	3,5	25	10
t/a at 8.000 h		29,500	29.000	48.000	2.000	11.000	4.000
appr. calorific value of waste	GJ/t	27,1	22,4	19,2	14,1	18,2	20,2
Year of Investment		1978	1976	1977	1978	1976	1975
Total Investment	Mio DM	19,80	27,8 [2]	25,0	1,60	6,023	1,50
Investment	DM/t	671	959	521	800	548	375
	DM/GJ	24,77	42,81	27,14	56,74	30,11	18,56
Electric Filter or Cyclone	Mio DM	-	0,405	1,2	0,095	-	-
Scrubber	Mio DM	2,70	0,993	1,6	0,175	0,580	0,370
Suction Blower	Mio DM	0,30	0,105	0,4	0,155	0,320	0,070
Reheating Flue Gases	Mio DM	0,25	0,598	0,5	-	-	-
Waste Water Treatment	Mio DM	0,15	0,473	n.a.	0,175	n. a.	0,010
Investment for Flue Gas Cleaning	Mio DM	3,40	2,574	3,7	0,600	0,900	0,450
	DM/t	115,25	88,76	77,08	300	81,82	112,50
related to total Investment	%	17,17	9,26	14,80	37,5	14,94	30,0
Operating Costs	Mio DM/a	3,20	9,07	8,57	0,900	3,535	0,700
Operating Costs [1]	DM/t	108,47	312,76	178,54	450	321,24	175
	DM/GJ	4,00	13,96	9,30	31,91	17,65	8,66
Electric Filter or Cyclone	Mio DM/a	-	0,272	0,32	0,022	-	
Scrubber	Mio DM/a	0,76	1,519	0,76	0,054	0,409	0,140
Suction Blower	Mio DM/a	0,38	0,043 [3]	0,25	0,0525	0,346	0,140
Reheating Flue Gases	Mio DM/a	0,08	0,508	0,75	-	-	-
Waste Water Treatment	Mio DM/a	0,06	0,233 [5]	0,06	0,0455	0,235	0,012
Operating Costs for Flue Gas Cleaning	Mio DM/a	1,28	2,575	2,14	0,174	0,986	0,152
	DM/t	43,40	88,79	44,58	87	96,92	38 [4]
related to Overall Operating Costs	%	40,00	28,39	26,97	19,33	19,58	21,71

1) Revenues for steam considered in case of heat recovery
2) including bunker (5 Mio DM)
3) of which 0,476 Mio DM/a are energy costs
4) no costs for neutralization
5) Oxidation of waste water and discharge of a COD-free waste water

currencies care should be taken to account for the differing rates for depreciation and interest on the capital invested, the various costs for repairs, servicing, maintenance, expenditure, personnel costs, energy costs and last but not least the different legal requirements for environmental protection.

8. CO-INCINERATION OF HAZARDOUS WASTE WITH MUNICIPAL WASTE

In certain, admittedly limited, cases it is possible to treat hazardous waste in domestic refuse incineration plants . Types of wastes which are predominantly involved are the following: waste oil, drilling and grinding emulsions, emulsions and mixtures of mineral oil products, mineral oils, mineral oil sludges, mineral oil refinery wastes, grease and wax wastes, varnish and paint sludges, solvents, solvent mixtures, sludges containing solvents, plastics and rubber wastes, cellulose production and processing wastes, pharmaceutical wastes, odorous food and spice processing wastes, and less toxic burnable industrial wastes.

The acceptance of such wastes for incineration is limited by various reasons:
- the possibility of boiler corrosion because of reducing atmosphere and acid flue gases;
- the possibility of exceeding thermal load capacity;
- mechanical problems (e. g.,with feeding and grate system);
- a lack of controlling device for incoming waste;
- difficulties with licensing authorities or local pressure groups;
- bad operational experience.

It may be stated that co-incineration of certain hazardous wastes with household waste may be a viable technical option for hazardous waste disposal under restricted circumstances. Future developments are likely to vary considerably: in some countries the idea of co-incineration may be promoted officially; in others it is unlikely to be promoted or may decline corresponding to a similar development in household incineration. Therefore,no general conclusion can be drawn at this stage.

9. HAZARDOUS WASTE AS A FUEL SUBSTITUTE

In times of rising fuel prices, the use of hazardous wastes as a fuel substitute is becoming an increasingly attractive proposition. In this context it is not only the cost aspect which is of significance: consideration should also be given to the fact that the oil which is thereby saved can be used as a raw material for other purposes. It is, in addition, possible that the noxious substances and residues produced during incineration may be completely bound in special production processes without detriment to the environment or to the products as in the case of the cement manufacturing industry. This will be discussed in greater detail in one of the following reports. The question is, whether there exist other industrial plants in which hazardous wastes are used on an operational basis as a fuel substitute. At present this occurs chiefly in the chemical industry where non-chlorinated hydrocarbons, waste oils and solvents are used.

This desirable development means that the calorific value of hazardous wastes, which are deliverd to off-plant incinerators (of the supra-regional type) sometimes tend to decrease. The waste generators use more and more of the better wastes in their own facilities as fuel substitutes and only rely on hazardous waste incinerators if the calorific value of waste is very low or if the wastes are very difficult to handle because of their toxicity. Thus the average calorific value of the delivered hazardous wastes would appear to be at present approximately 20 GJ/t.

Operators of supra-regional hazardous waste incineration plants have therefore suggested that government should become involved in the hazardous waste disposal route, in the extreme presumably establishing the obligatory use of certain incineration facilities, because it would not be logical to use primary energy, e. g.,auxiliary fuel, to burn the low calorific hazardous wastes. Such a demand seems to be rather short-sighted. Obviously generators of low calorific hazardous wastes have been 'subsidized' as long as their wastes were 'co-incinerated' with better wastes. If wastes with higher energy contents are omitted, the costs of treatment for the remaining wastes will rise. Following the polluter-pays-principle, the generators of such wastes would be charged with the real, higher disposal costs. This seems to be the best incentive for them to consider how to avoid or diminuish the generation of such hazardous wastes.

10. FUTURE PROSPECT

Many disposal techniques have been utilized or proposed, none of which is problem-free. Problem areas encountered have included adverse environmental impacts, excessively high costs and scarcity of acceptable sites.

As the quantity of hazardous wastes increase, the need for advanced waste management technologies will become increasingly important. Even if carried out under the strictest controls, landfill disposal is not effective for final disposition of certain types of toxic wastes. In this respect incineration as an integral part of the whole disposal concept is likely to be the best environmental option for final disposal of many organic hazardous wastes, provided the most stringend controls are exercised.

The future application of hazardous waste incineration will therefore be determined essentially by the following factors:
-   developments in the volume of waste generated, having regard to waste and waste avoidance;
-   development of alternative treatment technologies;
-   conditions placed on incineration on environmental protection grounds;
-   and thus development of costs;
-   better supervision of waste generators and waste streams through the control of statutory bodies;

-    potential citizen opposition to either incineration facilities or alternatives, such as land disposal.

ACKNOWLEDGEMENTS

The following companies have made available material to illustrate this report:
-    Claudius Peters Industrieanlagen GmbH, Hamburg
-    Bayer AG, Leverkusen
-    Kraftanlagen AG, Heidelberg
-    L. & C. Steinmüller GmbH, Gummersbach
-    Hessische Industriemüll GmbH, Wiesbaden
-    Lurgi Apparate-Technik GmbH, Frankfurt

REFERENCES

Disposal of Hazardous Wastes, Thermal Treatment, NATO/CCMS Report Number 118, Bruxelles, March 1981

Engineering Handbook for Hazardous Waste Incineration,
Draft, for U. S. Environmental Protection Agency,
Industrial Environmental Research Laboratory, Cincinnati, Ohio,
August 1980

Fabian, H. W.; Reher, P.; Schön, M.:
How Bayer incinerates wastes, Hydrocarbon Processing,
April 1979, pp. 183 - 192

Müll und Abfallbeseitigung;
Hrsg.: Straub, H.; Hösel, G. und Schenkel, W.;
Erich Schmidt Verlag Berlin

Working Group 'Rauchgaswäsche' :
Über die Rauchgaswäsche bei Rückstandsverbrennungsanlagen der chemischen Industrie,
Müll und Abfall 4/1980, pp. 101 - 108.

NORWEGIAN PLAN FOR DECENTRALIZED HAZARDOUS WASTE TREATMENT

Per Waage

Ministry of Environment
Oslo, Norway

SUMMARY

Norwegian industry has been closely involved in the solution
of the hazardous waste problems. A comprehensive plan has been
worked out by industry and the environmental authorities jointly.
Centralized treatment cannot be justified because of high investment costs in relation to the quantities to be treated. Decentralized hazardous waste treatment is envisaged utilizing as far as
possible existing plants and processes. State grants and low-
interest loans will be provided to develop and refine a decentralized system.

At the first stage 15 to 20 local collection sites are necessary. A central site for i.a. chemical analysis, preliminary handling
and packing for transportation to final treatment/disposal is required.
Larger quantities are to be transported directly from producer to
final treatment plants.

Strict control measures of the whole system is necessary.

Legal, economic and administrative measures to fulfil the plan
are examined, and the establishment of an administrative unit
(company) to operate and develope further the hazardous waste
system is envisaged.

INTRODUCTION

A 1975 Report to the Norwegian Parliament on measures to
control pollution stated that the State Authorities have a particular responsibility to contribute to effective collection, transport and treatment of hazardous waste. A preliminary scheme of 9
collection sites was envisaged. The investment costs were to be
covered by 2/3 state grants and 1/3 low-interest long-term loans.
The idea was to gain experience on quantities and types of
hazardous waste. By mid 1980 only one collection site had been
established and a couple of other sites were at the planning stage.
There are two main reasons for this rather sad situation:
Firstly, local protests regarding the establishing of collection
sites have led to great difficulties. Secondly, the lack of a
comprehensive scheme for hazardous waste handling has made the
municipalities uncertain whether the waste collected would be subject to final treatment. In addition, legistation covering
hazardous waste handling (registration, treatment, monitoring)
is still to be implemented.

During the last couple of years, private companies have shown
an increasing interest in handling their own waste as well as waste
from other producers. This is partly a consequence of environmental
requirements to abate pollution, and is partly also linked to increasing energy prices. Substantial quantities of organic hazardous
waste have become of interest as fuel. This is the reason why the
Norwegian cement industry at a factory in Slemmestad near Oslo has
undertaken tests on the incineration of pumpable organic hazardous
waste. The plant has also performed a series of analyses in order to
clarify the emissions and any possible effects on the quality of the
cement. The tests were conducted with financial support from the
Ministry of Petroleum and Energy and the Ministry of Environment.

The tests at Slemmestad resulted in a collaboration between
the environmental authorities and the cement industry with a view
to elucidating the possibilities of developing a nationwide system
for collection, transport and treatment of hazardous waste, where
the cement kilns would play an important role as a treatment unit.
The cement industry's interest was based on the assumption that
it will be an economically profitable proposition for them.

At the same time, alternative systems for collection, transport
and treatment of hazardous waste were to be investigated.

It has also been an assumption on the part of the Ministry of
Environment and the State Pollution Control Authority that the study
would include all types of hazardous waste, and that the system will
have to be organized in such a way as to make it possible to
follow and control the streams of waste from producer to treatment
plant.

THE WORKING METHODOLOGY

The Norwegian nationwide plan for hazardous waste handling
was prepared by a steering group consisting of members from
A.s. Norcem, the State Pollution Control Authority (SPCA) and the
Ministry of Environment (ME). To assist the steering group, four
expert groups were established: one group on collection and transportation, another on techniques and processes, a third group on
economic issues and a fourth on regulation and control. Representatives from Norcem, SPCA and ME participated in all four expert
groups. Representatives from the Federation of Norwegian Industries
joined the groups on collection and transportation and regulation
and control.

A reference group was also appointed, and it was consulted on a
continuous basis by the steering group. The following Ministries
and institutions were represented:

- Ministry of Industry
- Ministry of Social Affairs
- Ministry of Local government and Labour
- Ministry of Petroleum and Energy
- Ministry of Defence
- Central Bureau of Statistics
- Oslo Municipality
- One of the counties (Sør-Trøndelag)
- Federation of Norwegian Industries
- Federation of Norwegian Municipalities
- Labour Union for Municipal Workers
- Labour Union for Chemical Industry Workers

In addition several individual firms were consulted.

Within the framework of the planning process a number of
specific investigations were undertaken.

The Norwegian Institute for Transport Economics examined the
whole question of collection and transportation of hazardous waste,
including investments and operating costs. Location criteria for
sites(collection sites) for intermediate storage were also dealt with.

The Norwegian Insitute for Energy Technology examined the
potential for recovery of metals. The report concluded that
technology exists for recovery of most metals, but the profitability
of increased recovery is too low.

A consulting firm on environmental issues (I/S Miljøplan)
undertook a comparative study on various chemical-fixing methods.
The report concludes that treatment methods exist which are suited
for inorganic waste containing heavy metals.

The Norwegian Geological Survey has examined the possibilities for permanent storage in mountain shafts. The report stated that such disposal seems to be possible, but further investigations are necessary.

Chemcontrol (a Danish Company) has been engaged in examining a rotary kiln for solid organic waste.

A private consultant, Mr. Ib Knudsen, examined the facilities needed for permanent hazardous waste handling at one of the Norwegian cement plants.

All the investigations referred to have provided necessary input to the work of the expert groups and the steering group.

TYPES AND QUANTITIES OF HAZARDOUS WASTE IN NORWAY

The Norwegian plan accounts for 18 types of hazardous waste:

Oily waste:

1. Waste oil (mineral oils with < lo% water and sludge)
2. Spent oil emulsions (cutting emulsions and degreasing emulsions
3. Other oily waste (sludge, etc., from cleaning of tanks, oil-containing waste from ships, waste from oil and petrol separators)

Organic chemical waste:

4. PCB-containing waste (also PCT)
5. Wastes containing isocyanates (TDI, MDI)
6. Phenol- and formaldehyde-containing waste
7. Spent solvents containing halogens, sulphur and nitrogen
8. Other spent solvents (turpentine, white spirit, thinner, xylene, benzene, alcohols, ketones, aldehydes, etc.)
9. Organic waste containing halogens, sulphur and nitrogen
lo. Other organic waste (paint and varnish residues from distillation, organic acids and other residues of organic chemicals and products)

Inorganic chemical waste:

11. Mercury-containing waste (metallic mercury, mercury batteries, waste from dentistry, mercury-treated grain and remains from the treatment process).
12. Cyanide-containing waste (heat treatment salts, electro-plating baths)

13. Metal-containing sludge (dewatered sludge with at least
    3o% dry matter from electroplating, galvanizing and
    graphic processes)
14. Other inorganic waste (pickling liquors, galvanizing
    baths, graphic and photographic liquors, salts and other
    inorganic compounds containing metals, acids or alkalis)

Miscellaneous waste:

15. Waste containing organic metal - compounds   (tetra - ethyl
    lead, tetra - methyl lead, metal - alkyl)
16. Waste containing pesticides
17. Waste containing pharmaceuticals
18. Waste containing laboratory chemicals

There is also a category number 19 for waste that cannot be
classified under the other categories.

The number of waste sources varies from 2-3 for categories
4 and 6 to more than lo.ooo for categories 8, lo, 14 and 16. The
major sources have been identified. The plan also contains a survey
of the quantities of waste of each category generated annually and
an overview of the amounts disposed of in an environmentally
acceptable way Approximately 40 per cent of the total amount of
hazardous waste is at present collected and disposed of satisfactorily. The figure for inorganic waste is substantially higher
than for oily waste and organic-chemical waste. In volume,
uncollected oily waste accounts for approximately 90% of the waste
not disposed of satisfactorily.

The steering group has maintained that it is unsatisfactory
that a great quantity of hazardous waste is disposed of by unknown
methods. The purpose of establishing a comprehensive national
scheme for collection, transportation and treatment of hazardous
waste should be to ensure that a substantially higher proportion
of the total amounts of waste is handled safely from a health and
environmental point of view. Furthermore, the steering group has
stated that it is difficult to measure in monetary terms the costs
and benefits of a comprehensive system. It is assumed, however,
that the health - safety and environmental benefits exceed the
costs. In addition, improved organization of hazardous waste
handling will mean that more of it may be of interest in the
context of energy saving and/or materials recovery.

PRINCIPLES FOR FINAL TREATMENT OF HAZARDOUS WASTE

In principle, there are two alternative ways of organizing
final treatment of hazardous waste. One option is to establish
a central treatment plant, which as a minimum must include a

rotary kiln with smoke cleaning and possibilities for accepting
barrels and other containers, and facilities for chemical fixing
of inorganic waste. Such a facility entails advantages and
disadvantages.
Among the advantages the following can be mentioned:

- The main purpose of the plant will be waste treatment.
  This will be no subordinate activity as it would be in
  other plants where the main objective is production of
  goods.
- Most types of hazardous waste can be treated at one site.
- There will be adequate competence at one site for all
  aspects of handling of hazardous waste.
- Simple routines for acceptance and deliveries.
- Simple control and accounting (financial settlement).
- A central site can easily be expanded to handle larger
  quantities of waste.

There are , however, also clear disadvantages associated with
a central treatment site:

- High investment costs and associated relatively-high capital
  costs.
- A central site will require large land areas (12 to 25 acres).
- Possible difficulty to get local agreement for the location.
- A central site may entail higher transportation costs than a
  decentralized scheme.
- Final treatment of hazardous waste in a variety of existing
  industrial processes generally enables a higher degree of
  materials and energy recovery, as opposed to burning in a
  rotary kiln.

A number of plants in Norway exist which have invested in
facilities to handle their own waste and to some extent also waste
from other producers. In particular, we have a lot of facilities
to handle most of the inorganic chemical waste. This system can
be further developed and provide final treatment for most of our
hazardous waste. Investment in a centralized treatment plant can
not be justified on the basis of the quantities of hazardous waste
to be treated. By using existing plants and processes, we can benefit
from capital equipment already installed and expertise established
in various plants. Relatively moderate investments are required
to develop the system further in order to enable it to handle most
of the hazardous waste. Furthermore, a decentralized scheme is
already established and it avoids many of the problems faced by
building a central treatment site. In addition, plants can be
encouraged to treat on site or coordinate with other industries.
Cooperation between the authorities and industry can solve the
hazardous waste problems in a way which is optimal from a societal
point of view - by materials recovery and/or destruction.

For organic waste which can be pumped, cement kilns will, under certain conditions, play a major role. The test runs in this regard have been sucessful, and permanent incineration will start at one of the cement mills.

A new plant for handling organic chemical waste which cannot be pumped, is now being established. The waste will be dispersed in liquid and fed together with bark to a boiler furnace at one of our pulp and paper plants. Chemicals which might explode or particular toxic waste, according to the discharge licence, cannot be treated in this way. These may be fed into a cement kiln.

The system solutions for handling of inorganic and organic waste in a decentralized way after possible recycling has been carried out, are shown in diagram 1 and diagram 2.

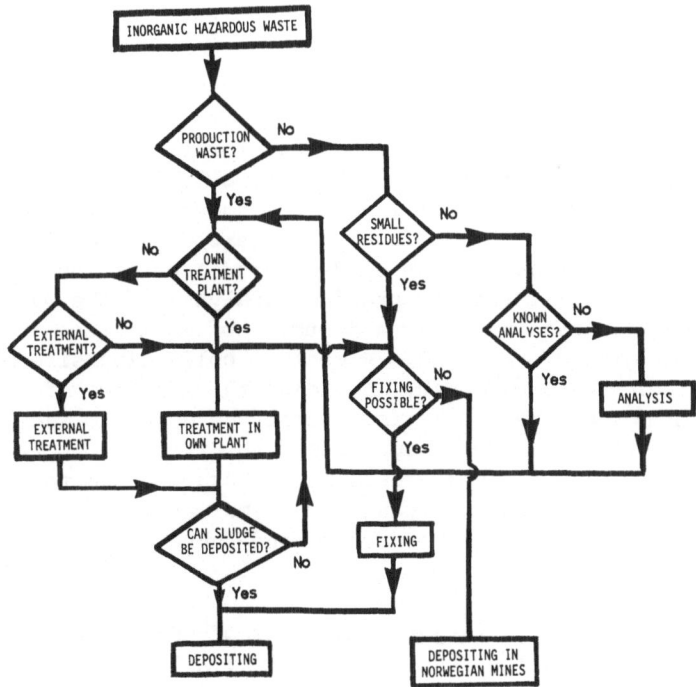

Diagram 1

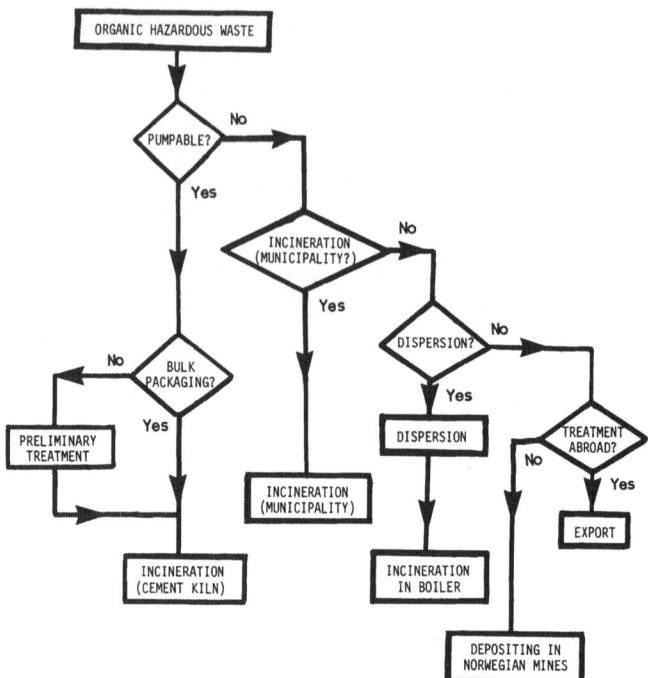

Diagram 2

COLLECTION AND TRANSPORTATION

When existing facilities are to be used for final treatment, hazardous waste will have to be channelled to a lot of different plants. This requires very careful sorting and channelling to a suitable treatment plant. A number of receivers of waste impose certain conditions as regards size of delivery, type of containers, etc. Some receivers will require proof of analysis of the waste. Export of waste will require a certain number of facilities for storage and preliminary treatment. A central collection site must be established for these purposes, including the handling of residuals which can not be treated satisfactorily by domestic facilities. The central site will have to be supplied with extensive expertise as regards analyses and further treatment of hazardous waste. Staff will be needed to transfer waste from smaller containers to stores/transport tanks, for re-packing and subsequent transport to final treatment.

To the extent possible larger quantities of hazardous waste should be delivered directly from producer to final treatment unit. The waste which cannot be handled in this way must be transported from producer to a local collection site. The system of collection sites will be established gradually. In order to take care of most of the waste 15-20 sites are necessary at the

first stage. The collection system will be gradually refined.
Municipal or intermunicipal waste handling facilities can be used
as location sites. It is assumed that the local collection sites
are to be equipped with containers which can be transported directly
to the central collection site, which has to analyse, classify and
repack the waste for transportation to recovery or destruction.
In addition to this kind of activity, the central site will be
responsible for information to the general public, municipalities
and plants on hazardous waste and education of personnel to operate
local collection sites.

MEASURES TO IMPLEMENT THE PLAN

     The Norwegian plan for hazardous waste handling considers
legal, economic and administrative measures.

a) Legal measures.

     The main principles for the management of hazardous waste
imply greatest possible freedom within a specific framework:

- Producers of waste are permitted to treat their own waste
  provided that this is done in a satisfactory manner
  (approved).
- Producers of waste are permitted to enter into direct agreement with the owners of approved treatment facilities for
  the waste in question.
- Receptor facilities are established and an associated
  arrangement for preliminary storage at a local collection
  site which will serve the producers who do not wish to treat
  their own waste and/or have too little waste to make a direct
  agreement with a plant for final treatment.
- The authorities will decide which substances and which quantities shall be subject to regulation at any time. In this
  way it will be possible to make priorities over time concerning what types of hazardous waste shall be made subject
  to legislation within the framework of the nationwide plan.

     It is proposed that the following regulation mechanisms shall
be used as a basis for further developement of the system:

1. General regulation (administrative provision) concerning
   hazardous waste.
2. Issue of permits to establishments with relatively large
   quanitites of hazardous waste, to the extent that such
   permits are not already used.
3. Approval arrangement for final treatment plants.
4. Approval arrangement for collection sites. In the first
   place a permit; subsequently provisions.

   5. Approval arrangement for transport of hazardous waste.
      Permits for the first enterprises, and afterwards provisions.

b) Economic instruments.

   In principle four types of economic incentives to control
hazardous waste can be undertaken:

   1. Incentives to reduce the generation of waste.
   2. Incentives to stimulate improved waste handling.
   3. Incentives to increase collection of waste.
   4. Incentives to improve the treatment possibilities.

   The steering group has concluded that general financial assistance to establish collection sites will be necessary to develop
an efficient system as soon as possible. Today's scheme of direct
state grants and low-interest loans for collection sites and treatment facilities will be extended. Selective economic assistance
to increase collection of specific types of waste should not be
introduced before experience with the regulatory actions and
collection sites are gained. The need for economic motivation is
determined by the effect of these regulatory measures. Economic
incentives could contribute to increased collection of hazardous
waste, but will at the same time require substantial administrative
efforts and may also be contradictory to the polluter-pays
principle. A transport subsidy should be examined as a part of the
implementation of the hazardous waste plan. Before new assistance
schemes to increase collection of waste are introduced, an assessment of the costs of collection and benefits gained should be
undertaken. Small amounts of hazardous waste may be costly to
collect and it should be clarified which environmental damages are
avoided by collection.

   Specific charges should not be introduced to finance economic
assistance. Such charges cannot be justified in light of the
amount of money needed, and may also be contradictory to general
budgetary principles. Economic assistance should be financed
by general state tax revenues.

c) Administrative measures.

   Apart from the functions to be undertaken at a central
collection site, the steering group recommended a central administrative body (company) with a governing board to coordinate the
operation of the nationwide hazardous waste scheme. The company
should undertake the following tasks:

   - Administer the operation of the whole hazardous waste scheme.

- Be owner of containers and necessary transportation equipment
  and operate the central collection site.
- Be owner of and operate future treatment facilities for
  hazardous waste which cannot be handled by existing plants.
  and processes, including possible storage in mine shafts.
- Be responsible for export and import of hazardous waste.
- Develop and refine the whole scheme in the light of experience
  gained.
- Be responsible for information and assistance to municipalities, collection sites and industrial plants.
- Control the operation of the system according to guidelines
  from environmental authorities.

The steering group recommended that the company should be
established as soon as possible to implement the scheme. Some
parts of the scheme could, however, be implemented without establishing a company, e.g., the licensing activity.

CONCLUSIONS

The steering group recommended that a nationwide scheme for
collection, transport and final treatment of hazardous waste
should be based on the following main elements:

- Utilizing as far as possible existing processes in the
  Norwegian industry for final treatment of hazardous waste
  in an economical and environmentally-acceptable way.
- A nationwide network of collection sites for hazardous
  waste, equipped at a minimum with closed containers for
  transport.
- A coordinated transport scheme for hazardous waste based on
  containers.
- A specific system for collection of oily waste, separated
  from the container system.
- Direct transport of larger quantities of hazardous waste
  from producer to final treatment.
- A central collection site for preliminary treatment of
  hazardous waste for subsequent transport to final treatment.
- An administrative unit (company) with operational responsibility, having its own management (board).
- A principle that final treatment facilities shall be available in Norway for all types of hazardous waste in Norway.
  However, it may be necessary to treat some hazardous waste
  in other countries. Likewise, Norway may import certain
  types of waste for final treatment.
- Economic assistance for necessary investments. Existing
  financial assistance schemes will be expanded.

- Regulation of the various elements of the system by means of administrative provisions and individual licences and control for all parts. It will be based on existing control and regulatory actions, supplemented with regulations in the new pollution control act.

ACKNOWLEDGEMENTS

This manuscript has benefitted substantially from extensive comments from Civil Engineer Bjørn Sveen, Federation of Norwegian Industries; Head of Division Richard Fort and Chief Engineer Magne Røed, Ministry of Environment; Assistant Director General Olav Nedenes, State Pollution Control Authority; Technical Director Ingvald Haga and Plant Manager Knut Trovaag, Norcem Inc. I also wish to acknowledge my debt to Mrs. Astrid Nilsen who typed the manuscript.

Finally, I wish to acknowledge all those who participated in the preparation of the Norwegian plan for hazardous waste handling.

HAZARDOUS WASTE INCINERATION IN A CEMENT KILN

Knut Trovaag

A.S. Norcem
Slemmestad, Norway

SUMMARY

Cement kilns can be used to incinerate organic waste and toxic substances. The energy from the waste is effectively used. The wet-process cement kiln is especially well suited since it essentially consists of a large high-temperature combustion chamber and has a built-in wet and dry scrubbing system capable of removing acidic gases. Solid particles are absorbed in the kiln or precipitated in the high efficiency dust precipitators and recirculated to the kiln. Provided appropriate operating control is maintained, the quality of cement is not affected.

The emission and imission levels are low and the destruction efficiency is very high compared to other incineration processes. The economy of using waste as fuel appears to be favourable both to the cement plant, the waste producer and the society.

INTRODUCTION

The cement industry is a typical energy consuming industry. The energy bill may be reduced by the reduction of energy consumption or by the use of low cost energy resources. It seems vital to the cement industry to pursue both alternatives.

The cement is produced either by wet or dry process. The wet process grinds and mixes raw materials such as limestone, sand and clay with roughly 3o - 4o% water. The rotary kiln receives these materials as a slurry. The dry process grinds and mixes the raw materials without addition of water.

Due to lower fuel consumption the dry process is in general more competitive than the dry process. However, about 3o% of the world cement production is still based on the wet process.

Capital requirements in the cement industry are heavy, both with respect to the conversion of wet process to dry process, the conversion from oil to coal, and the meeting of new air pollution standards.

The cement industry is not considered as an attractive industry by most investors today. Based on this situation we have to search for new possibilities in economical operations.

The wet process could still be competitive by developing new methods to reduce energy consumption, and through the incineration of waste fuels.

REDUCTION OF ENERGY CONSUMPTION

At Norcem's plant in Slemmestad, Norway, a program has been going on for several years resulting in a reduction of energy consumption.

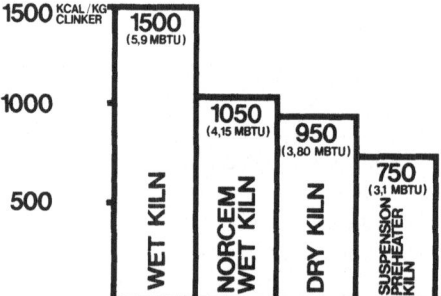

Figure 1.   Energy Requirements for Cement Kilns.

This reduction is achieved through modification of the chemical and physical properties of the slurry-feed to the kilns. The water concentration of the slurry-feed is reduced by adding surface-acting agents such as lignin chemicals, which is a waste from the bi-sulphite pulping industry. The slurry water content is now reduced from 33% to 26% and the energy consumption is reduced by approx. 2o%.

Further reduction in fuel costs may be achieved by the incineration of waste oil, solvents and hazardous waste. These waste materials create problems regarding deposition, or alterna-tively the need for building new high-cost incineration plants. A modern cement kiln has advanced environmental equipment fitted. The large high-temperature (above 14oo°C) combustion chamber ensures decomposition of stable materials such as polychlorinated biphenyls (PCB) and polyaromatic hydrocarbons (PAH). High efficiency preci-pitator units clean the stack gases from dust. The acidic gases produced are neutralized by the alkalinity of kiln materials. Stack gases, recycled dust and the clinker are controlled by chemical analysis.

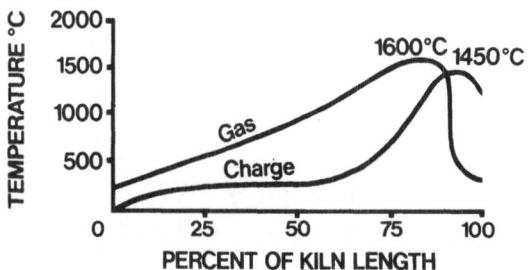

Figure 2.   Test Kiln Temperature Profile.

INCINERATION OF HAZARDOUS WASTE

Several cement plants have performed tests with waste incine-ration. Stringent tests have proved that there are no harmful effects to the environment as long as optimal burning conditions are maintained.

Reports from demonstrations in Canada and Sweden confirm the necessity of careful project planning in close cooperation with the Authorities and the employees. Early information to the plant neighbours and the newspapers is important.

Organization and planning

Two years ago NORCEM informed the Norwegian Ministry of Environment about the potential of utilizing cement kilns to dispose of organic waste. A project group was organized with representatives from the employees and the Authorities. Equipment specifications, emission and imission control programs were established. Detailed emergency plans were prepared to deal with potential accidents. The responsibility of waste transportation was assigned to the waste supplier. The fire brigade, the police, the medical doctors and our own rescue team established safety procedures for waste handling. Information to all our employees, to local authorities, to local health inspectors and to the newspapers was given. The Authorities supported a major part of the investment costs for the pilot plant.

Storage and handling

Variables to be considered in selecting equipment and handling methods include the waste's toxidity, flamability, corrosivity, polymerisation tendency, vapor pressure, viscosity, water and solid content.

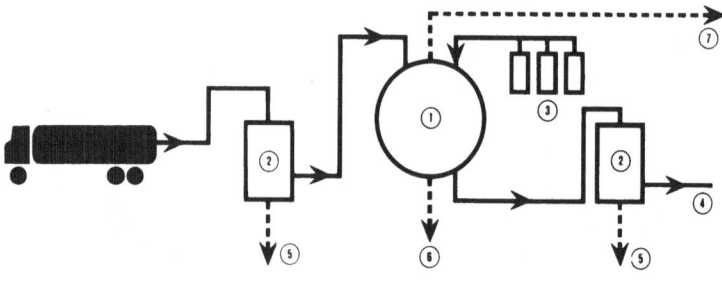

① **Storage tanks (3 x 20m³)** ⑤ **Solids to the clinker cooler**
② **Filter**   ⑥ **Water drainage to the slurry feeding**
③ **Inert gas**   ⑦ **Vapour to the kiln**
④ **Waste to the kiln**

Figure 3. Storage and Handling.

Three storage.tanks each of 2o m^3 are designed to the highest
safety standards. The tank vapor phase is ventilated to the kiln.
The tanks are pressurized by inert gas. Water drainage from the
tanks is connected to the kiln slurry feed system. Solid particles
from the filters will be fed to the clinker cooler. The storage
tanks, pumps and filters are installed into a concrete bassin.

Fire extinguishers, breathing masks, gas and smoke detectors
are located near to the storage tanks. Appropriate equipment and
safe operation procedures are necessary to obtain approval from
the Authorities and acceptance from the employees.

The test kiln

The test kiln is 155 meters long, and the diameter is 4 meters.
Normal fuel consumption is 92 tons of fuel oil for a production of
85o tons clinker a day. A waste burner is installed separately in
order to atomize the liquid waste by ultrasound air. The burner
tolerates solids up to a size of 5 x 5 mm. Interlock stops the
flow of waste fuel by kiln upsets. Complete incineration is
controlled by kiln temperatures and by monitoring oxygen and
carbonmonoxide concentration of exit gases.

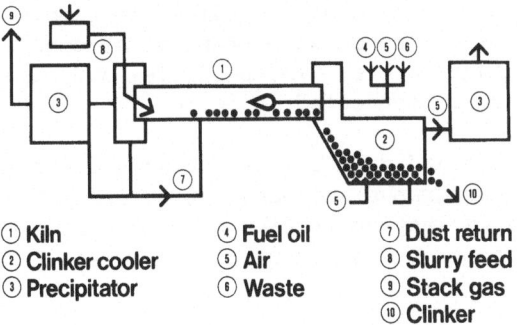

① Kiln   ④ Fuel oil   ⑦ Dust return
② Clinker cooler ⑤ Air   ⑧ Slurry feed
③ Precipitator   ⑥ Waste   ⑨ Stack gas
                                 ⑩ Clinker

Figure 4· The Test Kiln.

TEST AND CONTROL PROGRAM

Emission from a cement kiln using waste fuel will vary with
several factors such as chemical compositions of the waste, ratio
of waste to ordinary fuel, the kiln chemistry, the efficiency
of the precipitators and the dust return system. A special emission
and imission test program was arranged with the Norwegian institute
of Industrial Research (SI), the Norwegian Institute of Air Pollution
Research (NILU) and the Steam Users Association.

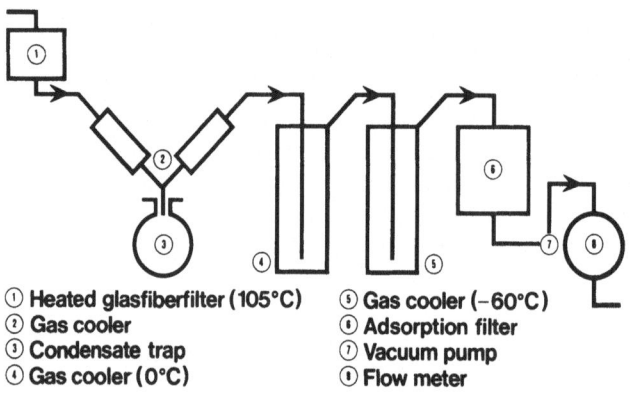

① Heated glasfiberfilter (105°C)   ⑤ Gas cooler (−60°C)
② Gas cooler   ⑥ Adsorption filter
③ Condensate trap   ⑦ Vacuum pump
④ Gas cooler (0°C)   ⑧ Flow meter

Figure 5.   Sampling Equipment.

Analysing for chemical traces in stack gases is complicated.
It requires handling and extracting large amounts of gas. The
gas sample is passed through a heated filter to remove particles
and then it flows through condensate traps at a temperature down
to -6o°C for collecting vapors. The particle- and vapor-
collecting functions are connected in series.

Vapor, condensate and dust from the kiln stack gases were
analysed by gaschromatographs and masspectrographs. Remarkable
advances in analytical instrumentation have made it possible to
determine contamination in concentration as low as (10^{-9} g/Nm^3).

Reference analysis

By incineration of ordinary fuel oil, the stack gas samples
were taken by normal burning conditions (oxygen > 1%) and by
unnormal burning conditions (oxygen < 1%).

The quantities of gas during all the tests were approx. 100.000 Nm^3/h (dry gas); 6o different organic components were detected. Table 1 shows some heavy organic components in the stack gas.

Table 1

	Normal operation		Innormal operation	
	μg/Nm^3	mg/h	μg/Nm^3	mg/h
C$_4$-Alkylbenzenes	1	9o	2	18o
Alcanes	14	127o	21	19oo
PAH	14	127o	25	227o
Thiophenes	7	635	13	118o
Ketones	6	545	16	145o
Aldehydes	4	365	12	1o9o

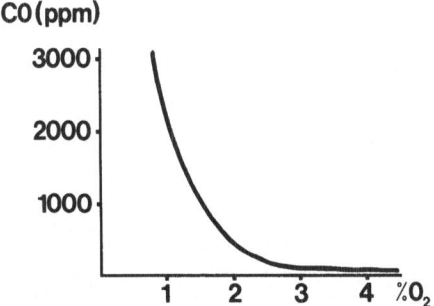

Figure 6.   Carbon-monoxide Concentration in Relation to the Excess of Air.

These results show the importance of constant monitoring of the excess of oxygen in order to achieve low emissions of organics.

Incineration of tar

The tar is a waste product from the cracking of natural gas. The main components of the tar are alkanes, alkylbenzenes and polynuclear aromatic hydrocarbons (PAH). The tar feed was 9oo kg/h.

7o different organic components were detected in the stack gas. Compared with the reference analysis special increase of aldehydes, alkanes and alkylbenzenes were found during the tar incineration. The feed of PAH was 123 kg/h. Based on this feed and the concentration of PAH in the exit gases, we found a destruction efficiency of 99.99999%. No organics were detected in the dust from the electrofilter or in the clinker.

Table 2. Stack gas emission of heavy organics.

	$\mu g/Nm^3$	mg/h	REFERENCE $\mu g/Nm^3$	mg/h
C_2-Alkylbenzenes	1o5	1o2oo	-	-
C_3-Alkylbenzenes	8o	776o	-	-
C_4-Alkylbenzenes	34	33oo	1	9o
Alkanes	15o	155oo	14	127o
PAH	14	136o	14	127o
Thiophenes	8	775	7	635
Ketones	23	223o	6	545
Aldehydes	96	931o	4	365
Phenol	2o	194o	-	-
Org. Acids	3o	291o	-	-
Alcohols	15	146o	-	-

Incineration of PCB (48% Cl)

During these tests the emission analysis were concentrated on heavy organic compounds and on chlorinated hydrocarbons. The feed of PCB was 5o kg/h.

Table 3. Stack gas emission of heavy organics.

	TEST I $\mu g/Nm^3$	mg/h	TEST II $\mu g/Nm^3$	mg/h	REFERENCE $\mu g/Nm^3$	mg/h
C_2-Alkylbenzenes	23	23oo	23	23oo	-	-
C_3-Alkylbenzenes	7	7oo	5	5oo	-	-
C_4-Alkylbenzenes	8	8oo	8	8oo	1	9o
Alkanes	48	48oo	54	54oo	14	127o
PAH	2	2oo	2	2oo	14	127o
Thiophenes	1	1oo	1	1oo	7	635
Ketones	4	4oo	7	7oo	6	545
Aldehydes	6	6oo	22	22oo	4	365

The difference between these tests and the reference could be a slight change in the burning conditions.

Table 4. Stack gas emission of chlorinated organics.

	Test 1		Test 2		REFERENCE	
	$\mu g/Nm^3$	mg/h	$\mu g/Nm^3$	mg/h	$\mu g/Nm^3$	mg/h
Chloroform	0.870	87	1.180	118	<0.00001	-
Carbontetrachlorid	0.400	40	1.220	122	<0.00001	-
Trichlorethylen	0.300	30	0.910	91	<0.00001	-
Tetrachlorethylen	0.400	40	1.210	121	0.00003	-
Chlorbenzenes	2.900	290	9.300	930	0.890	89
PCB	0.080	8	0.170	17	0.040	4

The analyses show an increase of chlorinated organic compounds in the stack gas. Destruction efficiency of PCB based on a feed of 50 kg/h is better than 99.9997%.

Polychlorinated dibenzodioxins (PCDD) and polychlorinated dibenzofurans (PCDF) are until now not detected. Special efforts are still made to identify these compounds.

Variable emissions of chlorinated organics could also be affected by chlorides in the water slurry to the kiln.

The following chloride concentrations were analysed during the PCB-incineration:

    Stack dust   max. 11.5% Cl
    Electrofilter dust max. 4.5% Cl
    Clinker   max. 0.02% Cl

Inorganic chlorides are concentrated in the stack dust and the electrofilter dust as alkali chlorides. An increase of stack dust emission from 5.2 kg/h to 6.8 kg/h was detected. The physical quality of the clinker was not affected but a slight increase of chlorid and a slight reduction of the alkali content was observed. No PCB was detected in the clinker or in the dust. Further incineration of chlorinated organics will be carried out.

Biological testing - Ames test

The emission analyses were supplemented by a biological test-ing. No increased canceriogenic activity in the stack was detected. An absolute conclusion from these tests is difficult to reach due to the influence of several unknown factors, such as toxic effects disturbing the analyses.

Imissions

Sulphurhexafluoride (SF_6) was sprayed into the stack gas;
26 stations for air sampling were located in the wind direction
during the tests. The maximum concentration was found to be
1:7o.ooo at a distance of 1 km from the stack.

Estimations based on wind measurements during one year indicates a max. hourly concentration of 1 $\mu g/m^3$ of air at a distance
of 1.5 km from the stack. The calculation is based on an
emission of 36oo g/h of gas or particles. The practical test
results and the results from the theoretical calculations are
comparable.

Summary of the tests

The test results are promising. We will, however, proceed
with further tests for another year. Incineration tests of waste
containing heavy metals, chlorinated organics, waste from the
VCM production and tar from the production of electrodes for the
aluminium industry will be performed.

The presence of chemical compounds in the stack gas will be
correlated with the fuel analysis and the process variables in
order to find the maximum safe feedding of hazardous waste.

All test results will be evaluated by the Authorities and
representatives from the employees before permission is given to
build a permanent plant.

However, we have started the technical/economical planning
for building a permanent system of waste incineration. The
technical principals of the permanent plant will be the same as
for the pilot plant.

ECONOMICS

The total investment costs for the pilot plant and the
control program is about o.9 mill. $.

The economy of burning hazardous waste in a cement kiln
depends on several factors. The fee charged for the disposal
varies with heating value, toxidity, water content, chloride-,
sediment content and the increased cost of clinker production.

The permanent plant will be based on the following
quantities of waste:

Waste oil	lo.ooo t/year
Solvents	4.ooo "
Chlorinated hydrocarbons	4.ooo "
Tar	2.ooo "
PCB	5o "
Total	2o.o5o t/year

Assumed an average heating value of 5.ooo kcal/kg. waste,
these quantities correspond to a saving of 16.ooo tons of coal
or lo.ooo tons of oil.

Investment costs

Engineering	5o.ooo $
Construction	2oo.ooo "
Mechanical equipment	1.5oo.ooo "
Electrical equipment	5oo.ooo "
Instruments and control equipment	15o.ooo "
Safety devices	loo.ooo "
Total	2.5oo.ooo $

| 6o% by the Authorities | 1.5oo.ooo $ |
| 4o% by NORCEM | 1.ooo.ooo $ |

Yearly operation costs

1 engineer	4o.ooo $
1 chemist	4o.ooo "
2 operators	5o.ooo "
Office	2o.ooo "
Maintenance (5 % of investment costs)	125.ooo "
Utilities	5o.ooo "
Overhead	4o.ooo "
Total	365.ooo $

Increased production costs

Kiln maintenance	5o.ooo $
Dust handling	5o.ooo "
Loss of earning:	
- 2o.ooo t clinker a 15 $	3oo.ooo "
Total	4oo.ooo $

Increased shut-down time for the maintenance of the kiln and the dumping of electrofilter dust due to increased concentration of inorganic chlorides is estimated to a production loss of 2o.ooo t/year.

Total yearly costs

Capital costs,	
- 2o% of Norcem's investment	2oo.ooo $
Operating costs	365.ooo "
Increased cement production costs	4oo.ooo "
Total	965.ooo $

Increased insurance liability costs are not included in the estimates. It seems likely that the burning of waste in a cement kiln will not require increased coverage.

Total yearly income

	Quantity t/year	Fee/ $	Total / $
Waste oil, 5o% water	1o.ooo	o	o
Solvents	4.ooo	9o	36o.ooo
Tar	2.ooo	9o	18o.ooo
Chlorinated hydrocarbons	4.ooo	1oo	4oo.ooo
PCB	5o	8oo	4o.ooo
	2o.o5o	-	98o.ooo

Cost/benefit analysis

Acceptable environmental solutions to a difficult disposal problem is to the benefit of the society.

An efficient low-investment solution to the incineration is beneficial both to the waste producers and the Authorities.

The saving of conventional fuel, such as coal and oil is to the benefit of the cement producer. By a cement production of 500.000 t/year, the savings of conventional fuels could be approx. 20%. The wet kiln using waste is competitive to a modern dry kiln using conventional fuels.

CONCLUSION

There are some fundamental laws of nature preventing us from carrying out chemical reactions with 100% efficiency. These limitations result in prohibitive increased costs as we strive for 100% conversion. Finding that economic optimum is what chemical engineering is all about. The economic optimum of several industrial projects is often based on the environment as a free depository. The environment can no longer be a free depository and this situation creates several new opportunities. The key to acceptable solutions is confidence and collaboration between industries, Authorities and the society.

A significant potential exists in using wet cement kilns to safely dispose of hazardous waste. Waste gypsum from the phosphoric acid production and fly ash from coal fired power stations are also used to mix into the cement by Norcem.

The cement industry is able to contribute to the elimination of waste problems with echological, economical and environmental acceptable solutions.

Here is the challenge and here is the opportunity for giving benefit to all of us in the future.

REFERENCES

- Complete reports from the tests

- Sentralinstituttet for industriell forskning
  Rapport nr. 8o o4 lo - 2
  15. juni 1981.

  Forsøk med destruksjon av flytende spesialavfall i en
  cementovn.

- NILU
  Referanse 2228o
  Juni 1981.

  Spredningsforsøk med sporstoff A.S. NORCEM, Slemmestad.

- NILU
  Referanse 2228o
  Juli 1981

  Vindoperasjoner og spredningsberegninger
  ved A.S. NORCEM, Slemmestad.

- NILU
  Referanse 21181
  Juni 1981

  Overslagsberegning av faregrenser ved brann i lagertanker
  for problemavfall.

- Kjelforeningen
  Målerapport 14. og 15. mai 1981
  Utslipp av støv og HCl.

INCINERATION AT SEA OF ORGANOHALOGEN WASTES:

TECHNICAL AND JUDICIAL MEANS TO CONTROL THE PROCESS

Jean-Marie Massin

Ministere de l'Environnement
Neuilly sur Seine Cedex, France

INTRODUCTION

In 1973, a new industrial process for thermal destruction of organo-chlorinated wastes at sea was submitted to the Ministry of the Environment.

This new process was based on the existence of special naval units having the following points in common :

- a variable volume on-board storage capacity.

- an incineration installation consisting of one or more tower furnaces, with variable dimensions, but all covered inside with refractories and fitted at the bottom with : 1) horizontal burners fired with atomized compressed air (fig.1), located on the periphery of the furnace(s) and facing towards the centre of the latter to enhance the gas turbulence inside the combustion chamber, and 2) a forced ventilation device providing the excess air needed to completely burn the products to be incinerated.

As far as their implementation is concerned, they do not have any fundamental differences, with actual incineration being performed in two phases :

* A preliminary phase during which the burners were started with fuel oil to bring the temperature of the walls (measured by means of thermocouples) to approwimately 1000 to 1400°C, depending on the type of installation,

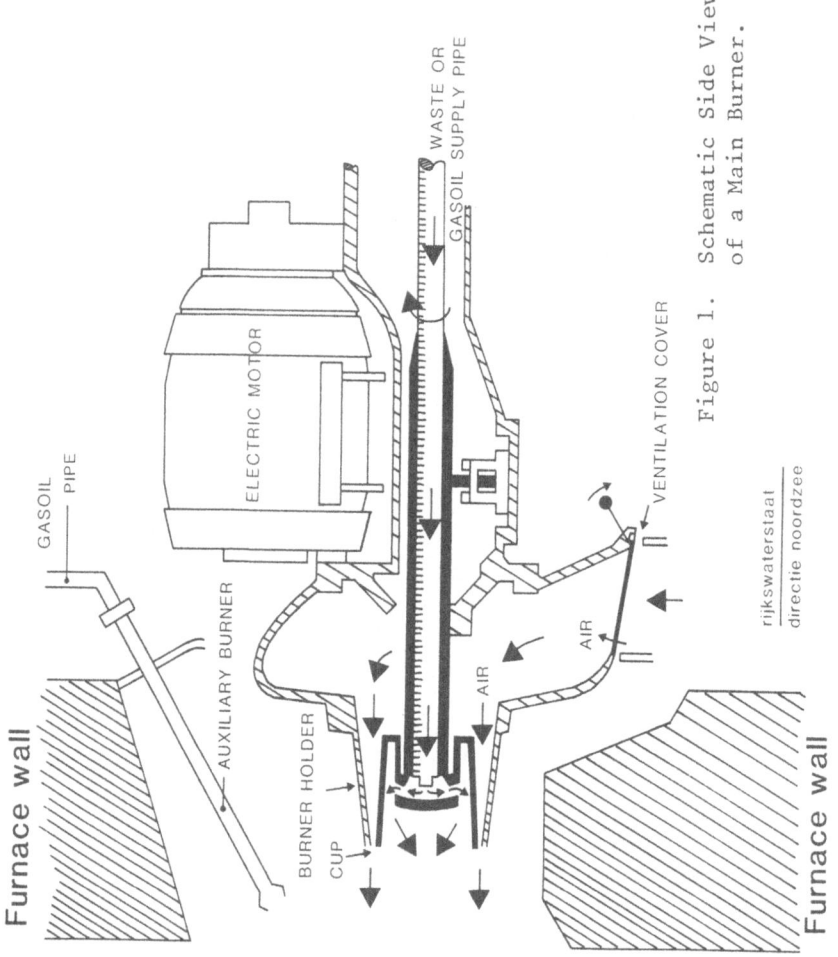

Figure 1.   Schematic Side View
of a Main Burner.

rijkswaterstaat
directie noordzee

* An actual incineration phase during which, once the required
temperature of the walls has been reached, the injection of
fuel oil was stopped and replaced by injection of the wastes
from one or more storage tanks, the flow of products to be
incinerated being adjusted to hold a steady-state temperature on the order of 1200 °C.

Implementation of the system foresaw that, if the heating power of the products to be incinerated proved to be too
low (less than 3000 kcal/kg) or if the temperature measured
by the thermocouples accidentally fell below 1000°C, an automatic system would cause reinjection of fuel oil.

Finally, since there was no combustion gas scrubbing
installation, all the gas was discharged to the atmosphere.

At the time, a preliminary examination of the file led
the relevant department of the Ministry of the Environment to
adopt a careful approach to the matter and to propose that,
before making any decisions on the new process, an in-depth
study would be made.

This approach answered a double finding :

- From the legal standpoint, incineration at sea operations were not apprehended by either international law
  or domestic law. Therefore, neither any International
  Convention nor, from the domestic viewpoint, any legal
  provisions could give the Administration the regulation tools enabling it to manage this type of activity.

- From the technical standpoint, the file was lacking a
  certain number of elements for assessment which would
  have enabled the pertinent authorities to judge the
  advisability of authorizing this new practice, as
  concerns principally the protection of the marine
  environment.

PRODUCTS PROPOSED FOR INCINERATION AT SEA

The products proposed for incineration at sea were waste
products mostly consisting of chlorinated hydrocarbons with
the general formula : $C_x H_y Cl_z O$, (in which x may equal 2, 3
or 4, y may equal 0, 1, 2, 3 or 4 and z may equal 2, 3, 4, 5
or 6) obtained by a reaction to add chlorine to aliphatic,
aromatic or heterocyclic hydrocarbons.

These products, basically coming from production lines
using chlorinated hydrocarbons as base products $\diagup$ production

of plastics (PVC), numerous pharmaceuticals, insecticides and
pesticides (DDT, PCB,....) and chlorinated solvents $\diagup$ posed
a real problem for a few years owing to :

- the large quantities involved, the production being nearly
  3.5 % of the production of the main product ;
- their toxicity with respect to living matter, their
  "tolerable" concentration limit in water often being less
  than one microgram per litre which is a concentration much
  below their solubility limit;
- finally, the stability of their molecular structure, causing a great persistence in the toxicity, even after dilution in the environment.

In order to cope with this problem, especially with
respect to environment protection, various onshore destruction possibilities existed on the industrial level.

Nevertheless, these wastes, in contrast to fuels, burn
poorly as a general rule; therefore, their incineration in
sophisticated installations poses numerous problems due, in
particular, to their low heat of combustion, their high
viscosity, the presence of solid particles and, in some cases,
polymerization or decomposition of the different components.

As far as environmental protection is concerned, it
should be pointed out that, in addition to being at a high
temperature, the burned gases resulting from this operation
are saturated with hydrochloric acid which must be removed
before the gases are dispersed into the atmosphere, the
tolerable concentration limit for the environment being
around 7 ng per m3 of air.

Some sophisticated installations that operate reliably
thanks to advanced technology make it possible to recover
this acid. Nevertheless, it is found that, except for re-use
on site, the recovered hydrochloric acid poses more problems
than it solves. Indeed, it is a product with considerable
over-production whose economic value is mediocre and unstable
and whose storage and transport also pose new, costly
problems.

Other installations neutralize the acid by scrubbing the
gases and treating the effluents with caustic soda and lime.
A new difficulty arises in this case involving the production
of a new waste (sodium or calcium chloride) which is practically non-toxic but produced in large quantities which,
once again, must be eliminated with consideration being given
to the current or upcoming regulations which prohibit all

discharge into water, rivers in particular being at the limit
of the tolerable level as far as their salt contents are
concerned.

THE FIRST RESEARCH CAMPAIGNS CONCERNING INCINERATION AT SEA

In this context, it seemed that the dispersal of the untreated combustion gases into the atmosphere, provided the
plume falls on the surface of the sea before reaching the in-
habited coasts, could meet a real need of the chemical industry, if the marine environment was not jeopardized.

In order to ensure this last point, the French Ministry
of the Environment, in close collaboration with the promoters
of incineration at sea, was responsible for setting up and
coordinating a full-scale experiment intended to give the
French Administration the elements for assessment it was lacking.

The first experiment was carried out in April 1974 off
Rotterdam in the area specially allocated to the incinerator
ships by the Dutch authorities. It was conducted with the
logistic support of the Institut Scientifique et Technique des
Pêches Maritimes, the Commissariat à l'Energie Atomique and
the Centre National pour l'Exploitation des Océans (CNEXO).

It was carried out based on both the special nature of
the products basically consisting of chlorinated hydrocarbons
and the characteristics of the incineration installations
existing on-board the incinerator ships VULCANUS and MATTHIAS
II.

Results

The results obtained from this experiment greatly corro-
borate the finding previously obtained in other countries
(especially Holland).

In particular, there was confirmation that the chlorinated compounds and any sulphur- and nitrate-containing compounds
present or detected in the initial products would no longer be
found in the gaseous effluents and that, consequently, all
organic -C Cl bond (generating toxicity) and brought in the
presence of excess air to a temperature above 900°C for more
than one second, was destroyed, with the production of simple
compounds recognized as being tolerated by the marine environment : nitrogen, carbon dioxide and monoxide, chlorine and
hydrochloric acid, excess oxygen.

In contrast, the same effluents were found to contain

new compounds that were not detected in the initial products
and having organic -C Cl bonds.

Two hypotheses were made in this respect :

* Incomplete destruction of certain components which presence
escaped analysis owing to the very low levels,

* Formation of new molecules from free radicals owing to the
existence in the furnaces of zones in which the ideal temperature and dwell time conditions for the molecules were
no longer brought together (case, for example, of the tops
of the furnaces when they are wide open to the exterior).

As for the influence of the plume "fallouts" on the
marine environment, the experiment demonstrated that introduction of the major combustion products into the environment
caused no disturbance; in particular, the acid "fallouts" did
not result in any pH changes greater than the sensitivity of
the instruments used, i.e., 0.05 pH units for 15 seconds.

Therefore, the influence of incineration on the physico-
chemistry of the marine environment almost instantaneously
neutralized the hydrochloric acid introduced.

Finally, from the ecological standpoint, no disequili-
brium was found in the zooplankton populations and no significant difference was detected in the composition of samples
taken both outside and inside the incineration area.

In conclusion

In conclusion, this experiment, which was conducted with
extensive logistics and analytical means, not only greatly
confirmed the validity of the process as an industrial process for the destruction of organo-chlorinated wastes, but
also demonstrated the potential risk that might exist for the
marine environment with discharge connected with poor occas-
ional or conceptual operation of the combustion furnaces, gas
containing highly toxic chlorinated products resulting from
incomplete destruction of the initial products and/or re-
structuring of new molecules from free radicals, the nature
of which cannot be precisely determined considering the
extreme complexity of the products proposed for incineration
as a result of the practically unlimited number of organic
molecules and their instability.

CONTROLLING INCINERATION AT SEA - ITS PRINCIPLES

These findings led the Administration, with technical

backing from the Commissariat à l'Energie Atomique, to seek
the control means enabling the authorities in charge to
express the discharges of effluents coming from the incineration units in terms of acceptability.

As far as the organo-chlorinated compounds are concerned,
their tolerance limit (which varies with their nature) is
nevertheless at a very low level : one milligram per m3 of
water. This special feature has two consequences :

- destruction must be performed at as high a level as
  possible to reduce the risk of extensive immersion;

- "direct" control of the increase in the concentration
  of organo-chlorinated compounds in the seawater (at
  level 10^{-9}) can only be made during "scientific" measurement campaigns based on long, delicate, expensive
  analysis methods (without mentioning the difficulties
  encountered in sampling the water since the "fallout"
  location constantly fluctuates owing to the wind).

This was the reason that led the authorities to seek a
routine "indirect" control based on the quality of the combustion gases leaving the incineration furnace, instead of
being based on the quality of the water at the plume "fallout" location or, yet again, based on the type of products
to be incinerated since they were found to be extremely complex and difficult to determine, even using the most sophisticated methods of analysis.

The first approach considered to achieve this purpose
was based on the finding that all the organic chlorine
injected as residues were found in the combustion gases mainly in the form of hydrochloric acid (HCl) and secondarily in
the form of chloride (Cl2) and organo-chlorinated compounds
that were not destroyed or were formed during incineration.
By establishing the " chlorine balance " in the combustion
gases, it is possible to reach the value of the ratio charac-
terizing the level of destruction, by assimilating the ratio
of the amounts of organo-chlorinated compounds with the ratio
of the amounts of chlorine they contain.

Then, the destruction level is expressed by the ratio
(in %) of the amount of chlorine in the form HCl and Cl2
(destroyed forms) which may be given by the expression
(Q + q') to the total amount of chlorine (injected form)
found both in the organic compounds (quantity q) and the
mineral compounds (quantities Q for HCl and q' for Cl2),
these quantities being measured simultaneously in the gases
leaving the furnace.

Therefore, this destruction level is expressed by the following formula :

$$To = \frac{Q + q'}{(Q + q') + q} \times 100$$

If these amounts are measured in the same sample of combustion gas, the concentrations of the various compounds in the combustion gas do not have to be known since their ratio is the only important thing : this does away with the difficulty inherent in measuring the volume of gases, a measurement which is always delicate, especially when the gases are corrosive and at high temperatures (near 1000 °C).

The analytical problem is thus well defined : metering, in the same gas sample, of the traces of organo-chlorinated compounds, hydrochloric acid and traces of chlorine, knowing that the ratio between the organo-chlorinated compounds and the hypochloric acid is around 10^{-4} and must be determined with a relative precision on the order of 5 to 10%.

The Commissariat à l'Energie Atomique was responsible for the development of an analytic procedure as well as instruments to implement this approach. Without entering into details underlying the reasoning having made it possible to estimate the optimum value for this ratio, it is nationally and internationally recognized that it must be above 99.9 % under all circumstances.

Although this is high for an industrial operation, the requirement of such an " efficiency " is realistic. Numerous measurements performed during industrial incineration campaigns have shown that a " well-run " furnace made it possible to obtain " routine " ratios of 99.95 % or even better (99.99 %).

Nevertheless, the measurement campaigns carried out on board the incinerator ships over the last few years have shown that, as a result of the complexity of the sampling and analysis devices, current technology makes it practically impossible to construct an instrument to measure the destruction level with a sufficient independent operating time to function in the absence of qualified personnel who cannot be present on board throughout the entire incineration campaigns (in particular owing to the very high expenses this entails).

Moreover, the method :

- can only be considered with chlorinated products,
- is an overall method that does not make it possible to

distinguish the various varieties of -C Cl molecules whose toxicity in relation to the marine environment can vary in considerable proportions.

Therefore, a second approach involving measurement of the combustion rate was proposed and retained for routine control in order to have a more " universal " control procedure for the time being better suited to the working restrictions on-board the ships.

Measuring the combustion rate on board the incinerator ships

If, indeed, we consider that the purpose of incineration is to use excess air combustion to " mineralize " the toxic organic molecules, another control method can be devised not based on measurement of the non-destroyed compounds by measuring the " chlorine balance ", but on measurement of the overall combustion quality.

This measurement may be apprehended by measuring the combustion rate Tc defined as the percentage ratio of the difference in the (CO2 - CO) content to the CO2 content in the combustion gases leaving the furnace. This measurement of the combustion rate has been traditionally used for years to characterize operation of fuel oil combustion units.

$$Tc = \frac{CO2 \text{ content } - CO \text{ content}}{CO2 \text{ content}} \times 100$$

This procedure was chosen by the American Administration (i.e. the Environmental Protection Agency when, in turn, it was confronted with the problem of incineration at sea in 1974 because of the request made by Shell-Chemical Co. located in Houston). It was experimented on board the Vulcanus (during incineration campaigns conducted in the Gulf of Mexico in 1974 and 1975) and on board a new unit, the Mattias III commissioned by the Hambourg shipyards in 1976 but decommissioned since (North Sea measurement campaigns, August 1976).

The practical advantage of measuring the combustion rate is to have well-known metering procedures and measuring instruments widely used in the industrial field and well suited to a high degree of automation so as not to require qualified personnel on board to operate them.

Nevertheless, although the procedure involving measurements of a combustion rate offers the advantage of being applicable regardless of the products proposed for incineration, it does have the risks of :

- not doing away with the possibility of forming new
  molecules in the parts of the furnace at which the temperatures or the excess air contents are too low ;

- only guaranteeing total destruction of the -C Cl molecules
  found in the wastes by throwing prejudice upon the homoge-
  neity of the temperature and the dwell times of the
  molecules in each part of the furnace.

Indeed, experience has shown that, under some circumstances (atmospheric conditions) and for special furnace con-
figurations (low gas inlet velocity associated with a large
furnace opening area), usually swirling movements of cold air
could lower the temperature of the combustion gases in the
top of the furnace, thereby resulting in incomplete combustion (or even reconstitution) of the organo-chlorinated compounds.

This dual finding actually poses the problem of the
comparative value of the measurement of the combustion rate
and that of the destruction level.

Indeed, under the above-mentioned conditions, although
the temperatures at the top of the furnace may be high enough
to pursue oxidation of CO into CO_2, they are too low to ensure complete combustion of the organo-chlorinated compounds
not yet destroyed (or formed) in the bottom and central parts
of the furnace. In this case there would be an increase in
the combustion rate at the top of the furnace (lowering in
the CO / CO_2 ratio) whereas the destruction level would
remain poor (presence of organo-chlorinated compounds in the
gases leaving the furnace).

In particular, this seems to be demonstrated by the
experiments conducted in France by the Physical Analytical
Chemistry Laboratory of the Ecole Polytechnique (Professeur
Guiochon) at the initiative of the Commissariat à l'Energie
Atomique. Indeed, the very first results seem to show that :

- there is considerable oxidation of CO into CO_2 for a few
  tenths of a second, as soon as the temperature exceeds
  750 to 800 °C;
- below 850°C, there is incomplete combustion of the
  selected compound (trichloethylene in this case) in
  equivalent dwell times.

Nevertheless, these results are only based on a single
series of experiments only corresponding to experimental
"laboratory" conditions differing from those of an industrial
furnace and taking combustion rates that are considerably
lower than the regulation values (85 to 99 % instead of

99.95 %).

It is still true that, since the possibility exists, it is important that :

* furnace construction and implementation be determined in order to prevent any possibility of introducing cold air at the top of the combustion chamber ;

* the combustion rate not be considered a priori as a ""universal " means of control (regardless of the furnace or the type of waste);

* the " real " temperature of the gases be measured at the furnace outlet.

Now that this remark has been made, it should be pointed out that, from among the additional conditions for implementing this procedure, there is the need to control the temperature in the furnace (greater than 1250 °C) and the oxygen content of the combustion gases (more than 3%).

The temperatures can be measured based on :

- direct flame temperature measurements carried out with optical devices (measurement of the thermal radiation by means of a pyrometer with reference to a calibration measurement) ;

- measurements of wall temperatures carried out using thermocouples regularly located in the walls to ensure constant coverage.

Experience has shown that, with a certain number of judiciously located thermocouples, it was possible to establish a correlation between these two types of measurements with a probability given in the following table for a flame temperature ranging between 1200 and 1300 °C.

Flame temperature	Wall temperature			
1200 °C 1300 °C	925 1140	915 1106	893 1067	838 1032
Probability	97.5 %	95 %	90 %	80 %

It still remains that, no matter what approach is taken, the control instruments must be fed with combustion gas. Thus, the gas sampling preblem is not solved and, in

particular, requires the presence of a technician on board to
ensure correct operation of the sampling device.

In order to remedy this situation, the Commissariat à
l'Energie Atomique has designed an " optical " instrument that
measures the emissivity in the infrared range of the gases
leaving the furnace to perform the measurement needed to
known which combustion gases there are while doing away
the constraints imposed by gas sampling operations.

Without entering into the theory underlying operation
of the optical measuring instrument and interpretation of
the results, it should be pointed out that the choice of a
passive measuring procedure (i.e. without exterior excitation
of the gases) makes it possible to collect the radiations
emitted by the gas molecules all along a diameter of the
furnace outlet section, thereby " averaging " sampling of the
information. In addition, the choice of the wavelengths makes
it possible to work both day and night.

By sighting the gases emitted by the furnace from a
distance of several meters (thus protecting the instrument
from the hot, corrosive plume), the instrument (called "PYRO
4") makes it possible to apprehend :

- the temperature of the combustion gases leaving the
  furnace;
- the partial CO and CO2 pressures (to compute the
  combustion rate);
- the concentration per unit weight of dust entrained
  by the gases.

After on-board calibration, all this information can be
used to constantly characterize the quality of combustion
throughout the duration of the offshore incineration campaigns.

This automatic instrument located far from the furnace
is designed for " unmanned " operation and recording of the
results in a black box.

JUDICIAL MEANS TO CONTROL INCINERATION AT SEA

Based on the above-mentioned results and the considerations stemming therefrom, incineration at sea problems have
been submitted to different international bodies:

On the worldwide level

The London Convention of 29 th December 1972 on
Pollution of the seas resulting from the dumping of wastes ;

On the regional level :

- The Oslo Convention of 15th February 1972 on Prevention
  of marine pollution by dumping operations performed
  from ships and aircrafts, signed by the States bordering
  on the Northeast Atlantic, the North Sea and the
  English channel and whose territorial scope is limited
  to the latitude of Greenland to the west and the longi-
  tude of Gibraltar to the south ;

- The Barcelona Cadre Convention of 16th February 1976,
  as part of the provisions set forth in the Protocol
  relating to the prevention of pollution of the
  Mediterranean Sea by dumping operations carried out
  from ships and aircraft.

If we disregard the special case of the Barcelona Convention during which the contracting parties quickly reached a
consensus of opinion concerning the prohibition of all incineration operations in the Mediterranean Sea, the London and
Oslo Conventions set up working groups which simultaneously led
to the establishment of a certain number of rules having an
authoritative nature, supplemented by technical provisions.

These rules lay down a certain number of principles, the
major ones being summarized as follows :

1- Offshore incineration is only a temporary method of
   destroying wastes and, therefore, must not in any way
   be interpreted as a natural state of things to dis-
   courage the search for preferable solutions as
   concerns the environment and the development of new
   techniques in particular.

2- Before proceeding with incineration at sea, the
   contracting parties must consider the practical
   possibilities of performing the operations onshore,
   using other methods of treatment, destruction or
   elimination, or of employing treatment procedures
   that reduce the harmfulness of the waste products.

3- Only a limited number of products are authorized for
   incineration at sea and they shall be specifically
   named or shall meet a certain number of criteria.

4- Each incineration system installed on incinerator
   ships (and, more generally speaking, the incinerator
   ships themselves) must first be certified under previously established strict technical conditions;
   certification is only to be valid for a set time and possibly renewed after it has been demonstrated that the
   ship continues to meet the standards initially set for

it.

5- All incineration at sea is subject to the issuance of a permit to incinerate and is only valid for an incineration system that has been previously certified.

6- Each incineration system must be subjected to a routine inspection procedure to make sure :

* the wall temperatures of the furnace(s) exceed 1200 °C;

* the minimum computed dwell time for the wastes in the combustion chamber, corresponding to a 1200 °C wall temperature, is on the order of one second ;

* the combustion rate is at least 99.9 % ;

* the destruction level at least equals the combustion rate, i.e., 99.9 % (measurement currently not compulsory owing to the lack of suitable measuring instruments).

7- All incineration must be carried out inside the special maritime areas whose limits meet a certain number of conditions including that of enabling the authorities in charge of controlling the incineration operations to intervene in situ without particular constraints in order to keep the users of the marine environment from incorrecly interpreting the exterior signs (flames or plumes) which might lead to course changes to provide assistance to a ship which appears to be in difficulty.

These areas are chosen by applying a certain number of criteria which bring the following aspects into play :

* biological : location of the areas in relation to spawning grounds, nurseries, fishing areas and shell fish areas ;

* occupancy of the maritime domain : location in relation to navigation areas, tourist sectors, mineral research or mining areas ;

* oceanological : currents, etc.;

* the characteristics of dispersal in the atmosphere of the area, especially the wind speed and direction the atmosphere stability, the frequency of inversions and mists, the types of precipitation and their extent, humidity. etc.;

* the oceanic dispersal characteristics in the area considered ;

* the existence of aids to navigation;

* the possible existence of underwater cables or pipes
  if the ship must drop anchor in the incineration
  area.

    The incineration industrialists have made some
proposals; they concern the use of transoceanic routes
to exercise their activity. These proposals have been
refused by all the contraction parties to the London
and Oslo Conventions.

This paper was presented by Mme. J. Aloisi de Larderel, of the
Ministere de l'Environnement.

HAZARDOUS WASTE

LANDFILL RESEARCH

Jack Bentley

Land Wastes Division
UK Department of the Environment
London, England

INTRODUCTION

Landfill practice within NATO countries has been reviewed in studies promoted by The NATO Committee on the Challenges of Modern Society.

The results of these studies which include details of landfill research were published in 1977 and 1981 (1) (2).

The final report presents the following conclusions and recommendations.

CONCLUSIONS

Landfilling can be a suitable method for disposal of certain types of hazardous wastes provided proper precautions in the siting, design, operation and post-operational care of the sites are taken to ensure adequate protection of the environment and public health on a long-term basis.

The results of landfill research can be used in the development of guidelines and codes of practice to promote improvements in the "state of the art" of landfill practice. The landfill pollution problem, however, remains complex and further research is required to improve the basic understanding of the attenuation mechanisms of leachate pollutants within and below landfills, and to develop means of preventing the ingress of water and the treatment of leachate. It will be necessary to develop cost-effective and environmentally sound control and prevention technology particularly for the treatment of leachates, the lining of landfill sites, and minimizing the generation of leachates.

Most of the landfill research has focussed on conventional inorganic pollutants while research on micro-pollutants such as trace organics has been more limited.

Research efforts to date have been directed mainly at the technical and scientific aspects of landfill problems while other aspects such as legal liability, site selection criteria, financial considerations for restoration and maintenance during and after active operation have been largely neglected.

International cooperation and exchange of information has been helpful in improving the understanding and knowledge of the problems associated with landfill operations.

RECOMMENDATIONS

Long-term environmental and health concerns should be thoroughly examined during consideration of landfilling as a technique for disposal of hazardous wastes.

Further scientific research should be conducted on the environmental impact of landfills as a disposal and co-disposal method for hazardous waste. In particular such research should examine the potential problems with micro pollutants (trace organics) and leachate collection and treatment.

Legal, economic, social and planning aspects of landfill disposal should be examined to improve the acceptability of the landfilling of hazardous wastes where water pollution problems are unlikely to occur.

International cooperation and exchange of information should be encouraged and continued to enhance the transfer of knowledge and to promote the development and implementation of sound landfill control and prevention technology.

The conclusions appear to find a broad measure of acceptance and attention is drawn to the proposition that landfill in the appropriate circumstances and under the right conditions provides an acceptable disposal option.

Though certain of the recommendations are the subject of debate there is little dispute regarding the need for additional research to rectify the present deficiencies in our knowledge of the subject.

LANDFILL RESEARCH

With the need for data relating to the disposal of hazardous
materials in landfill in mind, a review of the material available
was undertaken.

The following is a summary of what is considered to be the
most relevant information.

The United Kingdom government was one of the first bodies to
commission a systematic coordinated managed landfill research
programme and, in consequence of this, much of the material relates
to UK work.

HISTORY OF UK WORK

The report of The Technical Committee on the Disposal of
solid toxic wastes which was published in 1970(3) contained the
conclusion that insufficient scientific research had been carried
out on methods of disposal of toxic wastes.

Subsequent developments, in particular irregularities in
disposal practices and other events which led upto the Deposit
of Poisonous Wastes Act 1972, re-emphasised the need for knowledge
about the behaviour of toxic wastes deposited in landfill sites.
This need was reinforced by the requirement for the Department of
the Environment to offer guidance on the selection and control
of the disposal of hazardous wastes.

It was in this context that the Department commissioned a
large scale programme of research into the behaviour of hazardous
wastes in landfill sites. After a major survey of some 3000
sites in the UK in 1972 by an interdisciplinary team which
included geologists, hydrogeologists and chemists, it was concluded
that only 50 of the sites might present a possible hazard to a
major aquifer.

Nineteen sites were chosen from this list, taking into
account information from other sources, for a detailed investigation in the Landfill Research Programme over the period 1973 to
1977, with a final report of this work being published in 1978(4).

In phase II of the programme, significant pollution plumes
are still being monitored and other aspects of the programme are
continuing.

A major conclusion of the report is that sensible landfill is
realistic and an ultra cautious approach to the landfill of
hazardous and other types of waste is unjustified.

A similar conclusion was reached by the House of Lords Select Committee on Science and Technology after an intensive study of hazardous waste disposal in the UK.

In their report (5) they state,

"On the basis of present knowledge landfill is an acceptable disposal method for a wide range of wastes, provided that the suitability of the site is judged in each case. For many liquid industrial wastes codisposal with domestic refuse, if well executed, is also a valid method".

They also make the observation that,

"Research is needed to extend knowledge of the behaviour of wastes in landfill, with particular reference to advising those who have to decide locally on the suitability of sites for wastes; it is desirable eventually to classify wastes according to their suitability for landfill".

SUMMARY OF RESULTS OBTAINED FROM FIELD INVESTIGATIONS AND EXPERIMENTAL STUDIES

The fate of cyanide bearing wastes in the landfill environment has been studied experimentally and by field investigation. Initial laboratory experiments (6) suggested that hydrolysis of cyanide to ammonium formate, the release of gaseous hydrogen cyanide by the action of carbon dioxide and the reaction of cyanide with sulphur compounds to form thiocyanates were potential attenuating mechanisms. Controlled experiments, in which heat-treatment cyanide wastes in admixture with pulverised domestic wastes, at a loading of c. 0.005 kg CN/kg dry domestic waste, were leached in 2 m^3 cells under both aerobic and anaerobic conditions provided evidence for removal by loss as gaseous HCN, conversion to thiocyanates and conversion to other nitrogen compounds such as ammonia or nitrate. Overall, less than 3% of the added cyanide appeared in leachates over a three year experimental period. Destructive sampling of the cell after three years revealed not more than 3% of the original added cyanide remained and that the concentration within the cyanide layer had fallen from 48,000 mg/kg to 730 mg/kg.

Of the remaining 94% it was concluded that about half had been lost by volatilization of gaseous HCN and half as other soluble nitrogen compounds (4).

Field investigations were conducted on a site at Hammerwich, Staffs, where 50 tonnes of heat-treatment wastes, containing about 4 tonne cyanide, had been deposited in open drums in association with domestic and building trade wastes about 10 years prior to

the study. A maximum of 7.5 mg/kg of cyanide could be found in the
domestic waste with typical concentrations around 1 mg/kg.

Hydrogen cyanide gas was detected in boreholes drilled into
the wastes, indicating loss by volatilization. Thiocyanates were
detected in groundwater immediately underlying the disposal site,
whilst immobilization as relatively insoluble manganous ferrocyanide
was recorded. In all, less than 2% of the original 4 tonnes were
estimated to remain within or beneath the landfill.

Metal hydroxide sludge, rich in chromium and nickel, was
incorporated in another pilot-scale landfill cell. Leaching
during a three year period produced only 0.2% of Cr and Ni in the
leachate, at concentrations only a few times greater than those
derived from cells containing only domestic wastes. Evidence for
immobilization as sulphides was found in the anaerobic cells, but
exhumation of the sludges after three years showed them to be
essentially unchanged in composition and appearance.

The penetration of several heavy metal species under
unsaturated flow conditions through a Lower Greensand lysimeter
was studied using a synthetic leachate solution containing 100 mg/l
of each of the following ions: Cu, Sn, Pb, Zn, Cd, Ni and 10 mg/l
of Hg, together with appropriate concentrations of K (900 mg/l),
Na (1200 mg/l), Ca (500 mg/l) and Fe (100 mg/l), and the addition
of acetic, propionic and butyric acids.

These experiments, which have been extensively reported
elsewhere (7), showed that even after 600 days of irrigation, the
nickel, which penetrated most, had only travelled some 40 cms. It
did not migrate much further when leached for a further 200 days
with a solution of organic acids.

The attenuation of these species is therefore very strong in
a rock with a structure similar to the Lower Greensand and which
consists of a silica sand cemented in a calcareous matrix. The
approximate weights of Lower Greensand required to adsorb one
gram of metal were: lead 1.5 kg, chromium 2.7 kg, copper 1.4 kg,
zinc 4.5 kg, nickel 7.6 kg, cadmium 5.9 kg, mercury 6.2 kg.

Experiments have demonstrated that much of the retention is
associated with the precipitation of a ferric hydroxide phase
within the top few inches of the rock horizon, but other mechanisms
are involved.
Cadmium was shown, in the lysimeter experiments on the Lower
Greensand, to be very effectively retained in the top 40 cm of
rock.

Overall, lysimeter experiments, in which a metal-rich
synthetic leachate was irrigated onto the mineralogically complex

Lower Greensand aquifer showed that the majority of heavy metals
were retained by adsorption within the upper metre of rock.

Movement of metals through the Chalk and gravel aquifer
materials was found also to be slow, even though both contain
fewer clay minerals or hydrated oxides than the Lower Greensand.
Field investigations of the effects of the irrigation of sewage
effluents containing heavy metals onto the Triassic Sandstone
aquifer (8) have shown substantial decreases in concentrations of
chromium, zinc and to a lesser extent copper, within the upper few
metres of the unsaturated zone, the attenuation being ascribed to
adsorption. Site studies at landfills receiving metallic wastes
(4) also revealed significant attenuation between the concentration
in leachate in contact with the wastes and in adjacent groundwater.
For example, a landfill near Bromsgrove, on the Triassic Sandstone
aquifer was found to contain up to 1.7% zinc in solid waste samples,
but had concentrations of the order of only 0.1 - 0.5 mg Zn/l in
mobile water within the wastes, and did not appear to be making
any measurable addition to the natural zinc content of groundwater
beneath the site. However, significant movement of heavy metal
to groundwater was noted at a site at which large quantities of
liquid wastes of low pH were deposited on fissured sandstones,
and at another where sludges containing hexavalent chromium at
high concentrations were disposed of into the Triassic Sandstone.
In the former case, buffering of the leachate by the mineral in
the sandstone matrix and scavenging of heavy metals by the formation of complex hydrated oxides, reduced concentrations to low
levels within the oxygenated groundwater. At the latter site,
chromium in the anionic chromate form proved both persistent and
mobile, being attenuated only by dilution within groundwater
flows. Leaching cell experiments in which between 2 and 8 g/kg
dry domestic wastes of mineral oils were added showed that only
about 2% appeared in leachate and that the greater part of the
remainder was absorbed by the domestic wastes. Site investigations
at Rainham, Essex and at Wildmoor, Worcs (4) provided further
evidence of retention within the wastes and underlying aquifer
materials by absorption. At the latter site higher temperatures
in wastes and aquifer materials contaminated by oil suggested
biodegradation. High loadings of waste oils, in associations with
other liquid wastes may, however, lead to migration of oil through
the unsaturated zone with contamination of the water table, this
phenomena being found at sites in Scotland and Bedfordshire (4).

At Greenoakhill in Scotland, which is a co-disposal site,
the oil had been deposited in a number of small lagoons dug into
domestic waste. It was found that some 91% of the oil deposited
on the site was beneath the landfill above the water table.
8 per cent had moved to an area outside the area covered by
domestic waste and 1 per cent was found below the water table in
the vicinity of the site. Although much of the oil may have been

delivered as emulsions, it appears to have become immiscible with the leachate and to remain within 125 m of the site.

However, the main contaminant from the oil deposit that moved away from the site and which presented any hazard to groundwater was 43 mg/1 of phenol, associated with cutting oils, at boreholes outside the landfill. After a short distance, interstitial water in rocks contained less than 1 mg/l of phenol.

It has generally been found that the main water pollution problem from oily wastes is a phenolic contamination of the aquifer rather than an emulsified oily content or a separate oily layer. Where the sorption capacity of the domestic waste itself has been insufficient to retain the oil, usually because the local ratio of oil/refuse has been too high, oil will move to the rock/domestic waste interface. Provided the rock is not fissured the oil usually fails to penetrate as such into the rock, but remains in the algae and ferric hydroxide slimes at the interface. At lagoon sites this oily slime is frequently excavated so as to allow the penetration of aqueous leachate into the underlying rock structure.

The investigation of leachates from several sites containing industrial and/or domestic wastes showed that the concentration of phenols varied widely. The highest concentrations were recorded from the Eastfield site (270 mg/l) associated with liquid waste disposal, and from Maendy (60 mg/l) where mainly industrial wastes were deposited. At Ingham, high concentrations of phenol in landfill material (industrial solids/liquids) gave rise to concentrations of 390 mg/l in interstitial water in the Chalk 2 m beneath the site. Up to 15 mg/l were found in groundwater beneath the Hammerwich site where mainly domestic wastes were deposited along with builders rubble and a quantity of (phenolic) asphalt/tar, but no liquid wastes. It was not possible to relate concentrations of phenol in leachate to loadings at any site because of a lack of adequate records. However, from the above results it can be concluded that where phenolic wastes were deposited, some leaching of phenol occurred particularly where large volumes of liquid wastes were deposited.

At sites where phenols had been deposited the material was found to be moving towards the aquifer. However, at Ewelme (4) where the liquid centrifuged from chalk immediately below the domestic waste contained 400 mg/l of total phenol the concentration had fallen to 11 mg/l after passage of the leachate in the unsaturated zone through 15 m of chalk. This drop was more severe than the decrease in chloride ion caused by dilutional effects.

At Haigh Quarry (4) where phenol-containing liquids were deposited in a lagooning system which contained 8-100 mg/l of

phenol, with a mean for all liquids delivered of 7 mg/l, the mean
at the final lagoon in the series had fallen to 0.6 mg/l. This
was further diluted in a local river system.

At Flitwick (4) concentrations of 900 mg/1 in a lagoon had
fallen to 43 mg/1 a short distance down the water table.

At Eastfield Quarry (4), where the lagoon contained phenol
concentrations up to 20 mg/1, the boreholes some 40 m away, which
were connected by mine workings with the lagoon, had phenol
concentrations only a little above background.

The adsorption of phenol onto well-decomposed domestic waste
was shown to be reversible in the concentration range from
50-2000 mg/1. A higher fraction of the phenol is adsorbed at
lower concentrations and when fresh waste is used. Although the
adsorption factor is unlikely to be high enough to retain the
phenol effectively, the peak concentrations will be reduced
sufficiently to allow microbiological action to take place as the
more dilute solutions move through the waste.

The sand from the surface of the Uffington lysimeter (7) was
7 times more biologically active in decomposing phenol than sand
taken from a depth of 2.4 m, this reflects the reduced amount of
nutrient at depth below a soil horizon.

The high solubility and poor sorption of phenol by domestic
refuse in a landfill environment means that anaerobic biodegradation
offers potentially the most significant mechanism of concentration
reduction. The importance of biodegrading landfill material and
the unsaturated zone as bioreactive media for phenol decomposition
has recently been demonstrated.

The inhibition of phenol-degrading bacteria derived from
activated sludge has been observed at 250 mg phenol 1^{-1}. Under
experimental aerobic landfill conditions phenol degradation was
observed up to a concentration of 16,700 mg 1^{-1} in Lower Greensand,
and anaerobically at 8,200 mg phenol, 1^{-1} in a similar system (7).

The results are explained in terms of phenol degradation being
limited by diffusion of phenol across a stagnant liquid film
surrounding the microbes which are themselves attached to sand/clay
particles. This stagnant film is unlikely to be important in an
agitated aqueous suspension and the benefits of protection against
high inhibitory concentrations of substrate are unlikely to be
realised.

Due to the high solids content of landfill, a high percentage
of landfill bacteria are likely to be attached to solid supports,
particularly paper, and are likely to be protected from inhibitory

concentrations of phenols present in the aqueous phase. Microbes are also likely to be protected by sulphate or carbonate precipitates and by clay platelets derived from intermediate cover material.

The lowest initial phenol concentration used in the experimental work was 50 mg 1^{-1} and it was shown that initial concentration significantly affected the decomposition rate. At concentrations below 50 mg 1^{-1} the rate of decomposition will fall, but to an unknown extent. For instance it has been shown that little biodegradation of 2,4-D occurred when present at 2-3mg 1^{-1} although 60% conversion was obtained at higher concentrations. Possibly, the benefits offered by the solid phases at high phenol concentrations may prove disadvantageous for the complete metabolism of phenol. The final attenuating mechanism may be by dilution.

Small-scale tests indicated that phenol could be adsorbed by both aged and fresh domestic wastes which would delay the leaching of phenol from a landfill containing domestic wastes (10). Up to 100 mg/l of phenol were also degraded to methane and carbon dioxide by organisms present in a leachate from domestic wastes under anaerobic conditions (11). The maximum rate of degradation found was 129 mg/l week.

Drum-scale experiments which investigated the movement of phenol through aged (pulverised) domestic wastes (12) where the loading was 0.6 kg/m^2 (1.8 g phenol/kg dry domestic wastes) showed that when leached at a rate of 300 mm/y, phenol was reduced in concentration to background levels during percolation through 0.7 m of domestic wastes. Increased leaching of phenol (highest concentration 7 mg/l) was found where the infiltration of liquid was increased to 900 mm/y. At a high rate of infiltration of liquid (2500 mm/y) at this loading, a maximum of 450 mg/l was found in leachate, although 65% of phenol still degraded. A large contribution to leaching of phenol at the highest rate of infiltration was attributed to rapid leachate flow through larger channels within the wastes and possibly along the edge of the drums.

Model simulations indicated that the leaching of phenol at high rates of infiltration (c. 2500 mm/y) could be minimised by reducing both the loading of phenol (from 0.6 kg/m^2 to 0.06 kg/m^2) and the initial concentration of phenol leaching from the phenolic waste (12), although the simulations were not confirmed by experiment.

It was concluded that the concentration of phenol in leachate issuing from the base of a landfill would be related to the initial concentration of phenol in leachate from a phenolic waste, and on the 'residence time' of phenol within codeposited domestic waste beneath. The latter is a function of the rate of infiltration of liquids in the landfill, sorption of phenol, and the depth of underlying domestic wastes.

It is also possible that at high loadings (of readily leachable phenol: domestic waste) bacterial degradation of phenol would be inhibited or at least require some period of time for bacteria to acclimatise to high concentrations of phenol. Loadings of 0.6 kg/m^2 (1.8 g/kg) did not apparently inhibit bacterial utilisation of phenol in the drum-scale experiments.

Large-scale test cells, Edmonton

Phenolic lime mud (0.5% phenols) was emplaced above a 2 m depth of crude domestic wastes, at a loading of 2 kg/m^2 (2.6 g/kg dry domestic waste) (13). In 20 months of operation the maximum concentration of phenol in leachate was 206 mg/l (flow weighted mean 40 mg/l for the whole cell). Generally lower concentrations (maximum 90 mg/l, flow weighted mean 20 mg/l) were found for the central areas of each cell. The highest concentrations of phenol were found during period of heavy infiltration, particularly in leachates from the edges of the cell. Background concentrations of phenol in leachate from domestic waste only were a maximum of 10 mg/l (flow weighted mean 1 mg/l).

Model simulations indicated that at a loading of phenol of 2.6 g/kg dry domestic waste, and with a rate of infiltration of 300 mm/y through 2 m of domestic wastes, minimal leaching of phenol should occur, if the rates of degradation were similar to those found in the laboratory experiments (11, 12). Results from the Edmonton cell suggest that although degradation of phenol may be taking place within the cell, (0.1% of the phenol present has been leached) the extent of phenol biodegradation within the crude domestic wastes is less than that found in the laboratory-scale experiments, where pulverised domestic wastes were used.

— — — — —

In addition to the general landfill research, considerable specific work has been commissioned in support of advisory documents published in the Department of the Environment's Waste Management Series.

This included work on the co-disposal of acid wastes with municipal waste.

Here results from laboratory column experiments have shown clearly that fresh refuse is not suitable for acid co-disposal on account of the high risk of groundwater pollution. This is on the basis of:-

(i)   The low density of freshly received refuse encourages
        rapid fissure flow of acid through the waste. The same
        would also apply to poorly compacted mature refuse.

(ii)   The pH of leachates from a fresh fill may already be -pH
        5.0. It is unwise to reduce this still further by the
        application of acid.

(iii)   Application of acid to fresh refuse resulted in very high
        concentrations of iron and sulphate in the leachate, plus
        increased concentrations of heavy metals.

Use of visibly mature pulverised refuse (6 months old) with
a compaction density of 0.9 kg m^{-3} resulted in no acid break-
through for loading rates less than 124 ml/kg (= 7.2g acid/kg) for
the hydrochloric acid waste and 66 ml/kg (= 10.5g acid/kg) for the
sulphuric acid. No marked peak of iron or sulphate was observed
in the leachate in either case.

The most significant problem appears to be increased
leaching of heavy metals. The problem may be accentuated by the
suggestion that microbial activity is inhibited in the presence of
high concentrations of acid (> 10g acid/kg of refuse). Failure to
achieve anaerobic conditions will limit the precipitation of heavy
metal sulphides. The presence of a high concentration of sulphate
in the waste is not thought to present a residue problem if construction work is undertaken at a later date.

Summarising this it can be said that co-disposal of pickling
wastes containing less than 6% acid and 29% iron sulphate at
loading rates of 7.2g acid/kg of refuse and/or 124 ml liquid/kg
is acceptable provided that the wastes are deposited in a trench
dug in mature, well compacted refuse, with sufficient depth of
fill (at least 5 metres) to permit attenuation of the waste. The
additional point must be added that although the mobilisation of
heavy metals from pulverised waste was relatively low, many
pulveriser plants include a ferrous metal abstraction system.
Crude waste is likely to contain a much higher concentration of
metallic objects and the results of the field experiments are
required to evaluate whether increased concentrations of heavy
metals are produced under these circumstances. Neutralisation
prior to deposition may therefore be necessary in some cases.
- - - - -

Investigations were also carried out on arsenic bearing
materials. Results from experimental work on the incorporation of
arsenical waste in domestic waste landfill suggest that between 60%
and 90% of an applied charge of arsenic (III) or arsenic (V) in
solution, at concentration of 200 mg kg^{-1} and 2000 mg kg^{-1},
is adsorbed within 24 hours by both fresh and aged domestic
wastes. In general for arsenic (III) it seems that the higher the

pH value the greater the adsorption this phenomenon being more
evident for aged waste than for fresh waste. The converse seems
to be the case for arsenic (V). Samples of leachate from a domestic
waste landfill in the UK gave the following concentrations of
arsenic:

	Arsenic concentration (mg litre^{-1})
January 1974	0.4
March 1974	0.3
July 1974	0.15
June 1975	0.02
May 1976	0.02

Extensive tests have been carried out to determine the amount of
arsenic which might be leached from pfa by rainwater: measured
concentrations in leachate were low, about 1 mg litre^{-1} of arsenic
or less, and independent of bed depth up to 14.5m of pfa. These
results were obtained under conditions which closely simulated a
landfill situation.

A study was carried out on a site where calcium arsenate is
landfilled as a sludge containing 6% of arsenic with other wastes
from an inorganic chemical activity. Calcium orthoarsenate
$Ca_3(AsO_4)_2$ has a water solubility of 40 to 140 mg litre^{-1} @ 20°C
(ie 15 to 50 mg litre^{-1} as arsenic) and is produced in an impure
form from solutions containing arsenate or arsenite ions by
adding slaked lime and an iron salt. In addition the technical
precipitate contains secondary calcium arsenate $CaHAsO_43H_2O$, and
basic salts such as $Ca_5(AsO_3)_2$. It is believed that calcium
orthoarsenate can dissociate to form the more soluble (3,500 mg
litre^{-1}) secondary calcium arsenate, but that a high pH would
depress this dissociation, favouring the relatively insoluble
orthoarsenate. Leachate from this landfill accepting calcium
orthoarsenate typically contains arsenic at a concentration of
less than 5 mg litre^{-1}: all leachate is collected and returned to
an effluent treatment plant for treatment. Other work involved
the investigation of the fate of primary and secondary Batteries,
Pesticides and Beached Oil.

Attenuation Factors in Landfill

The mechanisms for the attenuation of wastes in moving away
from landfill sites have been largely identified but their quantitative assessment in terms of 'attenuation capacity' and rate at
which they occur still requires considerable study.

None of the research sites so far examined is considered a
significant risk to any groundwater resource or existing groundwater abstraction, but with the exception of Hooton (4) which is
essentially a containment site, each site gives rise to a pollution
plume of not insignificant size, which has required a surprising
number of boreholes for reasonable identification. In addition,
assessment of 'leachate-rock' interactions or rates of degradation
are generally difficult to obtain from field investigations, unless
the geology and hydrogeology are known in great detail and an
efficient monitoring network installed for long term measurements
of the plume movement which, inter alia, may result from seasonal
changes in leachage production, or from different methods of
landfilling or character of waste deposited.

The main conclusion is that heavy metals are often strongly
attenuated either by 'sorption' or possibly precipitation as
sulphides in reducing conditions invariably present in intergranular
aquifers. However, Eastfield Quarry (4) has illustrated that under
adverse conditions, (fissured, non-interactive aquifer), considerable movement of heavy metals may take place with no neutralisation and little opportunity for biodegradation of the leachate.

Biodegradation does not appear to occur by aerobic processes
in any of the unsaturated zones studied but may take place
anearobically. (Little attenuation is suggested at Ruckley (4).)
The oxygen rich groundwater at Hooton (4) appears to allow 90%
of TOC present in the unsaturated zone to be reduced by mechanisms
other than dilution. Where oxygen is depleted and malenclaves of
reduced groundwater are produced, then organic degradation in the
saturated zone may be very much less rapid. Zones of differing
redox conditions, e.g., Villa Farm (4), may also allow indigenous
components of the aquifer such as Mn and Fe to enter solution and
effectively add to the dissolved solids content of the groundwater.

Oils tend not to flow very far from landfills since they are
physically absorbed by intergranular materials, or on fracture
planes in fissured strata. However, where wide conduits are
present, e.g., Eastfield Quarry (4), considerable movement may take
place.

In all cases studied, chloride is unattenuated except by dilution and dispersion and forms a useful reference for assessing   ˙
the effects of chemical and biological attenuation mechanisms on
other waste components.

A general conclusion that can be drawn from the above work is
that although modern tip management techniques, including
compaction, disposal in bunded cells and the use of low permeability
intermediate covering materials graded to encourage run off, may

minimise the rate of leachage formation, it is probable that a
certain volume of polluted liquid will be formed in most UK
landfill situations. In its raw form the leachate is unsuitable
for discharge to surface water or, very often, for treatment at
sewage works. However, recent studies (9) have indicated that
simple forms of on-site aerobic biological treatment for organic-
rich leachates may be effective, particularly when additions of
phosphorus are made to correct the microbial nutrient balance.

In addition to attenuation in and below landfills, a further
mechanism which can mitigate the effect of pollution plumes is
dilution in groundwater. The potential effect of such dilution
was specifically examined in connection with study undertaken on
arsenic and the treatment employed is given by way of example.

TABLE 1

Aquifer	Dilution factor (50m site)	Dilution factor (300m site)
Chalk	0.4×10^{-2} to 0.1×10^{-1}	0.2×10^{-1} to 0.7×10^{-1}
Sandstone	0.2×10^{-1} to 0.7×10^{-1}	1.0×10^{-1} to 3.0×10^{-1}
Gravel	0.2×10^{-2} to 0.4×10^{-2}	0.1×10^{-1} to 0.2×10^{-1}

If the presence of arsenic-bearing waste in the landfill (a) as
calcium arsenate and (b) as arsenic trisulphide is assumed, and
that the leachate is a saturated solution with respect to these
materials, then the concentration of dissolved arsenic in the
leachate will be (a) 50 mg litre^{-1} or (b) 0.3 mg litre^{-1}. The
resultant concentrations of arsenic in groundwater may be
estimated as (in mg As litre^{-1}):

i.   The potential risks to groundwater from the landfill of arsenic-bearing wastes may be examined using mathematical modelling techniques developed as part of the Landfill Research Programme (14) enabling dilution factors (concentration of pollutant in groundwater: concentration in leachate) to be estimated for various types of aquifer. Under natural groundwater flow conditions in the aquifer, the dilution factors are proportional to the length of the landfill in the direction of groundwater flow, and to the net infiltration of liquids into the landfill. Assuming a typical annual rainfall of 1,000 mm and net infiltration of 300 mm, appropriate dilution factors are: See Table 1.

ii.   If a borehole is located directly down the groundwater gradient from a landfill, then all of the leachate derived from the landfill could be induced to flow to the borehole, if the pumping rate remains sufficient to cause diversion of groundwater flows over the front along which the leachate moves. So long as these conditions are maintained, the dilution factors (concentration in borehole discharge: concentration in leachate) depend only on the borehole abstraction rate, the leachate percolation rate and the surface area of the landfill. Using the same assumptions as in (i) dilution factors and resultant arsenic concentrations at various abstraction rates are (in mg As litre^{-1}). See Table 3.

iii. It must be stressed that these calculations deliberately represent a 'worst possible case' situation. In practice there are other factors which would tend to reduce the groundwater concentrations. We have assumed, for example, that the whole area of the landfill site contributes to the production of leachate, a situation unlikely in practice. The calculation also assumes that all the arsenic leached from the calcium arsenate entered the groundwater and that none is retained within the landfill, for example by chemical precipitation or absorption; neither does the calculation allow for attenuation within an unsaturated zone, but assumes hydraulic continuity between the landfill and an aquifer. Finally we assumed the high arsenic concentration in leachate from calcium arsenate of 50 mg litre^{-1}.

TABLE 2.

Aquifer	Calcium arsenate		Arsenic trisulphide	
	50m	300m	50m	300m
Chalk	0.2 to 0.5	1.0 to 3.5	1.2×10^{-3} to 3.0×10^{-3}	6.0×10^{-3} to 2.1×10^{-2}
Sandstone	1.0 to 3.5	5.0 to 15	6.0×10^{-3} to 2.1×10^{-2}	3.0×10^{-2} to 9.0×10^{-2}
Gravel	0.1 to 0.2	0.5 to 1.0	6.0×10^{-4} to 1.2×10^{-3}	3.0×10^{-3} to 6.0×10^{-3}

TABLE 3

Borehole discharge rate ($m^3 \times 10^3$ day^{-1})	Dilution factors		Calcium arsenate		Arsenic trisulphide	
	50 x 50m	300 x 300m	50 x 50m	300 x 300m	50 x 50m	300 x 300m
2.27	0.9×10^{-3}	0.3×10^{-1}	4.5×10^{-2}	1.5	0.27×10^{-3}	0.9×10^{-2}
4.55	0.5×10^{-3}	0.2×10^{-1}	2.5×10^{-2}	1.0	0.15×10^{-3}	0.6×10^{-2}
9.09	0.2×10^{-3}	0.8×10^{-2}	1.0×10^{-2}	0.4	0.6×10^{-4}	0.24×10^{-2}
22.73	0.9×10^{-4}	0.3×10^{-2}	4.5×10^{-3}	0.15	0.27×10^{-4}	0.9×10^{-3}

Future work

Apart from the continuing managed landfill research programme
involving long term monitoring of a wide range existing situations
and the work commissioned in response to specific needs, more
general work includes an examination of the viability of creating
artificial unsaturated zones under landfill.

This work was undertaken on the basis that research carried
out to date has identified the potential of the unsaturated zone
below a landfill for the attenuation of a range of hazardous and
toxic wastes. Fine grained materials, in which flow is intergranular, have been found to be particularly effective for the
attenuation of leachates by a number of physical, chemical and
biological processes. Detailed study of these mechanisms has
been undertaken using large in-situ and smaller 'monolith'
lysimeters together with laboratory experiments. Complementary
investigations of existing hazardous waste disposal sites were
also undertaken.

Particular attention has been paid to the attenuation of
heavy metals (Ni, Cd, Zn, Cr, Pb, Fe). Undisturbed lysimeters
from important aquifer formations have been irrigated with a
synthetic leachate containing heavy, alkali and alkaline earth
metal chlorides in carboxylic acid solution; short chain
carboxylic acids being an intermediate in the degradation of a
range of celluosic compounds, commonly accounting for up to 80%
of the total organic carbon in landfill leachates. The distribution of heavy metals in the lysimeters, following extended periods
of irrigation, has been determined by sequential selective chemical
extraction, employing increasingly aggressive reagents. The
geochemical fractions with which the heavy metals are associated
have been identified allowing some elucidation of the attenuation
mechanisms.

The importance of the unsaturated zone, has prompted the study
of a range of readily available fine grained waste materials,
which could be used to increase the thickness and therefore the
attenuation capacity of the strata beneath landfill sites, or to
create an unsaturated zone where one is absent. The function of
these materials is twofold:-

a. To chemically interact with hazardous components of the leachate,
   leading to their attenuation.

b. To form a physical barrier to waste migration, thus controlling
   their release into the natural groundwater environment.

Potential liners already identified include pulverised fuel ash (PFA) from coal fired electricity power stations, spent foundry sand a n d (formulated without the use of phenolic resins) fine limestone wastes from quarrying, and sand waste from china clay extraction. Overburden often removed in sand and gravel extraction, which may contain sand and silts with a high proportion of smectite clays (with a high ion exchange capacity) has also been considered.

Laboratory work initially to define the mineralogical composition of these materials was recently commenced, and anaerobic columns set up to irrigate these potential liner materials with a heavy metal leachate under conditions which associated research has shown to exist beneath landfill sites.

Using similar extraction techniques to those already outlined the distribution of the heavy metals in the liners will be investigated, subsequent to irrigation, in order to study the dominant mechanisms for attenuation.

Following successful laboratory work, a small-scale field trial is envisaged in which waste containing heavy metal sludges is deposited on a liner material previously instrumented for in situ monitoring of temperatures, interstitial pore fluids and gas composition. Drilling will subsequently be carried out to obtain contaminated samples of liner for analysis.

A major use for permeable liners is also to create an intergranular unsaturated zone in potential landfill sites in fissured strata in which leachate flow may be rapid. The intergranular liner will slow down the release of leachate into the aquifer such that natural processes, e.g., dilution dispersion, biodegradation etc, are sufficient to reduce the concentration of contaminants to acceptable levels before reaching usable resources.

In this way quarries in fissured strata, in need of restoration, may be reclaimed by controlled landfilling after engineering with a permeable liner.

Landfill Research Outside the UK

A literature search to cover relevant work on laboratory, lysimeter and field scale experiments produced well over 100 references for consideration. Not one reference covered direct experimental work on the attenuation properties of domestic refuse, and no data calculated by indirect means could be found.

The most relevant US reference (15), from July 1976, concluded that attentuation within refuse would depend on the particle size of the refuse, and that heavy metal attenuation

was affected by the formation of unknown complexes. This has
been deduced as a corollary of their work on attenuation within
soils. Although it is five years since this report, research in
the US has focussed on initial leaching, and attenuation in soil
and rock structures beneath the landfill site, in a similar
fashion to parallel work in Europe.

Several "sandwich" co-disposal experiments, similar to the
pesticide columns at the Harwell Waste Research Unit in the UK
have been set up using columns or lysimeters (16)(17)(18).
However, all these experiments have only examined leachate from
the base of the apparatus, with no reference to leachate immediately
below the waste layer, typically electroplating sludges or battery
waste, and therefore there is no information given on the attenuation provided by the refuse. Analysis of the waste layers before
and after the experiments carried out by the WRC (18) may have
showed that some metals are scavenged from the leachate percolating
through the waste layer, whilst the release of others to the refuse
below is increased. However, discrepancies in sampling were cited
as a more likely explanation for the variation observed in the
waste composition, and no other data from within the lysimeter is
available.

The other area of research which might have yielded information
concerns experiments on controlled leachate recycling. Two major
US projects reported in 1980 involved the codisposal of domestic
refuse with electroplating sludge (19)(20). Leachate was again
only sampled at the base of the lysimeters, and although leachate
concentrations levelled off, correlation with the quantity of
refuse present is not possible. Recent work in Germany (21)
followed a similar pattern, including experiments with intimately
mixed electroplating sludge and mature refuse. Leaching tests
were performed on the sludge, and correlation of leachate
concentrations of copper, chromium and nickel with the quantity
of sludge was observed. The intimate mixing of the waste does
however complicate any estimate of the attenuation to be expected
through depths of domestic refuse.

Specific work on arsenic wastes in Germany and the Netherlands
gave the following results.

At a site in West Germany field tests on arsenic/lead sludges
mixed with domestic waste have been made. The ratio of industrial
waste: domestic waste varied between 1:1 and 1:8, and it must be
inferred from the summary provided that other industrial waste
might also have been present in the test cell along with the arsenic/
lead sludge. Each test cell was 25 x 25 metres, filled with 5
layers, each of which was 2 metres thick, giving 200 tonnes of
waste per cell. Test cells were arranged in pairs and one was

compacted whilst the other was loose filled. The experiment ran from about 1969/70 to 1974 at least, and the following results were obtained:

Type of landfill	Loose fill	Compacted
Arsenic in leachate $(mg\ litre^{-1})$	0.65 average	0.17 average
	1.7 maximum	1.4 maximum

Data from the Netherlands shows the arsenic concentrations in leachate from domestic waste landfills to be 0.10 mg litre^{-1}. The waste was heavily compacted (to 900 kgm^{-3}).

Acknowledgements

The author is grateful to colleagues both within the Department and in its Research Contractors for assistance and contributions particularly, MR R G D OSMOND AND MR D E BOND (DOE), DR C P YOUNG AND DR C BARBER (WATER RESEARCH CENTRE) DR P J YOUNG (WASTE RESEARCH UNIT, HARWELL)

References

1. Disposal of Hazardous Wastes: Landfill No 64
    Committee on the Challenges of Modern Society NATO
    May 1977.

2. Disposal of Hazardous Wastes: Landfill No 120
    Committee on the Challenges of Modern Society NATO
    Feb 1981.

3. Ministry of Housing and Local Government
    Technical Committee on the Disposal of Solid Toxic Wastes
    Report of the Technical Committee /Chairman Dr A Key/
    HMSO 1970 (ISBN 11 7502782).

4. Department of the Environment. Cooperative Programme of
    Research on the Behaviour of Hazardous Wastes in Landfill
    Sites. Final Report of the Policy Review Committee. 169
    pages, HMSO, 1978.

5. House of Lords Select Committee on Science and Technology
    Hazardous Waste Disposal Report Vol I HMSO 1981.

6. MONTGOMERY, H A C BROWN, B L REDHEAD, D L and HOUGHTON, E A
    Study of Hazardous Wastes in Landfill Sites: Removal Mechanisms
    for Cyanide. WRC Report No 1303, Water Research Centre,
    Stevenage. 12 pages, 8 figures, June 1974.

7. **Uffington Lysimeters Operation and Results (Parts 1 to 4)**
    WLR Technical Notes 36, 40, 42 and 60 DOE London 1976, 1977,
    1978, 1979 Part 5 UKAE Harwell May 1981 HMSO.

8. Water Research Centre. Effects of Effluent Recharge on the
    Major Aquifers. Final Report to the Department of the
    Environment on Research to March 31, 1980. 15 pages, 3
    appendices, Water Research Centre Report LR 1179, 1980.

9. ROBINSON, H D Leachate from Domestic Waste: Studies of Aerobic
    Biological Treatment. Part 1, Batch Aeration Studies. WRC
    Technical Report – TR 135. 37 pages, May 1980.

10. Study of Landfill Disposal of Acid Tars and Phenol-Bearing
    Lime Sludges II. WLR Technical Note Series No 52 DOE London
    1977.

11. The Biodegradation of Phenol under Anaerobic Conditions by
    Organisms Present in Leachate from Domestic Waste WLR
    Technical Note Series No 63 DOE London 1978.

12. Study of Landfil Disposal of Acid Tars and Phenol-Bearing
    Lime Sludges III. WLR Technical Note Series 71 DOE London
    1980.

13. Behaviour of Wastes in Landfill sites. Study of the Leaching
    of Selected Industrial Wastes in Large-Scale Test Cells,
    Edmonton N London. WLR Technical Note Series No 69 DOE
    London 1980.

14. Dilution of Landfill Leachates in Groundwater. WLR Technical
    Note Series No 53 DOE London 1976.

15. Folsom, B L, Brannon, J M, Green, A J Residual Management by
    Land Disposal, Hazardous Waste Research Symposium Proceedings,
    USEPA. EPA 600-9-76-015, pp 86-93, July 1976.

16. Van Veen, F Research on Co-disposal in the Netherlands.
    SVA-3447-051. January 1980.

17. Banerji, S K (Editor). Management of Gas and Leachate in
    Landfills, 3rd Annual Municipal Solid Waste Research
    Symposium Proceedings. EPA 600-9-77-026, September 1977.

18. Newton, J R Pilot-Scale Studies of the Leaching of Industrial
    Wastes in Simulated Landfills. Water Pollution Control 1977,
    pp 468-480.

19. Kinman, R N, Walsh, J J Leachate from Municipal and Industrial
    Waste Landfill Simulations. Disposal of Hazardous Waste.
    6th Annual Research Symposium Proceedings. EPA 600-9-80-010
    pp 203-232, March 1980.

20. Pohland, F G, Gould, J P Stabilization at Municipal Landfills
    containing Industrial Wastes. Disposal of Hazardous Waste -
    6th Annual Research Symposium Proceedings. EPA 600-9-80-010
    pp 242-253, March 1980.

21. Stegmann, R Criteria for the Codisposal of Municipal and
    Industrial Wastes. 5th European Sewage in Refuse Symposium
    EAS Munich 22nd-26th June 1981.

HAZARDOUS WASTE LANDFILL RESEARCH: U.S. EPA PROGRAM

Norbert B. Schomaker

U.S. Environmental Protection Agency
Cincinnati, Ohio, United States

BACKGROUND

Congress, through the Solid Waste Disposal Act (also known
as the Resource Conservation and Recovery Act RCRA, PL 94-580),
directed the Environmental Protection Agency to develop programs
for the management of hazardous and non-hazardous solid wastes.
To effect the degree of coordination required of such an
endeavor, a committee was formed to prepare and update an Agency
wide research strategy. The Solid and Hazardous Waste Research
Committee coordinates the efforts of the Office of Research and
Development (ORD) with those of the Office of Solid Waste (OSW),
the Office of Enforcement (OE), the Office of Water Programs
(OWP), the Office of Emergency and Remedial Response (OERR), and
the EPA Regional Offices.

The Fiscal Year (FY) 1980 Solid and Hazardous Waste Research
Strategy was based on the anticipated needs of the Agency as it
implemented the mandates of RCRA. The major emphasis was placed
on hazardous waste with continuing efforts in the municipal
solid waste area. The FY 1981 strategy focused entirely on
hazardous wastes.

Agency responsibilities under RCRA are reasonably specific,
and future activities to fulfill those responsibilities have
been scheduled according to a step-by-step strategy. Within
this hazardous waste strategy, four research categories, which
are described in the following sections, have been established.
The four categories are:

o   Hazardous Waste Identification
o   Hazardous Waste Technology
o   Hazardous Waste Risk Assessment
o   Energy and Mineral Wastes

Hazardous Waste Identification

Hazardous waste identification includes sampling, analysis,
characterization, measurement, monitoring, and quality assurance
of measurements. The development of valid identification
protocols is vital to the separation of hazardous from
non-hazardous wastes and to the subsequent "cradle-to-grave"
management of hazardous wastes. At present, our knowledge of
many of the factors in the effective management of hazardous
wastes is far from complete.

The strategy for hazardous waste identification is based on
the need to provide a complete capability for the sampling and
analysis of wastes which are potentially hazardous. Initial
efforts will develop engineering and chemical screening methods
and protocols. Multimedia monitoring methods are equally
important and occupy a high priority in this area. Quality
assurance for the monitoring program, necessary to ensure the
validity of all of the data to be collected for characterization
and subsequent monitoring of hazardous waste sources, are
addressed.

This strategy was initiated in FY 1980 and will continue to
receive emphasis through FY 1985.

Hazardous Waste Technology

With the environmentally unsafe options for management of
hazardous waste eliminated by RCRA, generators must turn to the
use of properly designed and constructed landfills or other
alternatives. Improved treatment and disposal technology is
necessary if hazardous materials are to be managed in a manner
that is economically as well as environmentally sound.

The thorough study of all hazardous waste streams and
potential treatment and disposal technologies is not feasible
because of the variety of wastes considered hazardous; and funds
would not likely be available to conduct such a mammoth
undertaking. Immediate short-term efforts will focus on those
areas where the need is acute and industry research cannot be
expected to be sufficient to meet needs. The land disposal
strategy is presented in greater detail in following sections of
this paper.

EPA treatment research strategy is directed in three basic areas: preprocessing or predisposal treatment, thermal destruction and detoxification treatment. Short-term research efforts will concentrate on the refinement of existing technology for each of the three areas. Investigation of advanced technology, particularly in the area of thermal destruction will continue through FY 1985.

Hazardous Waste Risk Assessment

The paramount objective of EPA in the management of solid and hazardous wastes is the protection of public health from the effects of these substances by direct contact or through transmission through the environment. In order to accomplish this objective, it is essential that reliable tools be available to enable EPA to gauge the potential impacts of various substances under various circumstances.

The research strategy for risk assessment includes the production of risk assessment documents for specific chemicals (scientific assessment), determination of toxic effects on humans likely to result from exposure to hazardous wastes, development of methods and data to define environmental transport and fate characteristics of hazardous wastes, determination of likely ecological effects of environmental insult from hazardous wastes, and development of socioeconomic criteria and model scenarios of the impacts to be expected for various waste management incidents or disposal methods.

Research efforts in the scientific assessment and health effects areas will continue in succeeding years. The other areas will be initiated in FY 1982 and will also continue throughout the program.

Energy and Mineral Wastes

Certain industrial and mining operations yield large volumes of waste containing some hazardous constituents. These special wastes include uranium mining wastes; phosphate mining, milling, and processing wastes; certain other mining wastes; cement kiln dust; utility wastes (coal ash and flue gas desulfurization wastes); and gas and oil drilling muds and brines.

Specific efforts underway through 1981 include studies of wastes from coal-fired electric generating plants, uranium mining, phosphate mining and benefication, and from selected ore mining industries. The research program for FY 1981 and beyond also includes assessments of gas and oil drilling operations. Lysimeter studies of oil shale leaching will also be conducted.

The above four categories of research are being pursued by various research groups within the USEPA. The hazardous waste landfill research program is part of the hazardous waste technology category. Since the focus of this paper is orientated primarily towards hazardous waste land disposal research, it is this area which will be specifically addressed in this paper.

LAND DISPOSAL

The hazardous waste land disposal research program, encompassing state-of-the-art documents, laboratory analysis, bench and pilot studies, and full-scale field verification studies is at various stages of implementation.

The objective of this program is to compile a more adequate data base of information so that current hazardous waste disposal technology to the land may be upgraded by (a) developing proper site selection design and operational criteria for the establishment of new waste disposal sites, and (b) developing proper control technology for upgrading existing waste disposal sites. The procedures thus developed will consider and describe those specific functions required to eliminate the pollution potential from landfills, land treatment facilities, and surface impoundments. Specific functions include the ability to predict or provide guidance for leachate quality and quantity, quick indicator tests for the selection of lining materials, quick indicator tests for chemical fixation performance and compatibility with various generic waste streams, a methodology for minimizing moisture infiltration, alternatives to conventional landfills, and predictors of environmental/economic impacts.

Over the next 5 years the research will be reported as criteria and guidance documents for user communities. The land disposal research program is currently developing and compiling a data base for use in the development of guidelines and standards for waste residual disposal to the land as mandated by the "Resource Conservation and Recovery Act of 1976" (RCRA). Permit Writer's Guidance Manuals, which provide guidance for conducting the review and evaluation of permit applications, are currently being prepared. Technical Resource Documents (TRD) in support of the Guidance Manuals are also being prepared in specific areas to provide current technologies and methods for evaluating the performance of the applicant's design. The information and guidance presented in these manuals will constitute a suggested approach for review and evaluation and should be useful references for design and operation.

The current hazardous waste land disposal research program has been divided into five general areas: (a) Landfill ; (b) Land Treatment; (c) Surface Impoundment; (d) Economic Assessment of Hazardous Waste Disposal Practices and Alternatives, and (e) Technical Resource Documents (TRD) and Guidance Manuals.

LANDFILL

The landfill program examines components and unit operations of a hazardous waste landfill and optimizes their performance. The goal is to control and/or be able to predict the movement of liquids and gases in and around a landfill. The program develops basic information for each component and prepares reports of research and user manuals for permit writers, design engineers, and operators.

The components and unit operations of a landfill to be addressed in this research include:

o   Waste Leaching
o   Pollutant Movement
o   Pollutant Control
o   Pollutant Treatment
o   Waste Modification

Waste Leaching

The overall objective of this research activity is to provide techniques for leaching a waste sample to obtain leachate similar to that which will occur when the waste is landfilled. A leaching procedure is being developed for use in determining which pollutants will be of concern in movement of leachates out of the landfill and for evaluating leachate treatment schemes in order to better develop pollutant control technology. A Technical Resource Document (TRD) will be completed in FY 1982 and a field validation study will be implemented in FY 1983. The initial product will focus on leacate from well defined mixtures or single waste streams.

Pollutant Movement

The objective of pollutant movement work is to develop predictive relationships which can be used in the design and regulation of disposal facilities for municipal and hazardous solid waste to achieve safe release to the environment.

Because the processes or pollutant retention by soils do not operate independently of one another, understanding the causes of pollutant movement/retention in a specific situation is

difficult. Heterogeneity of soil characteristics and the usual
lack of knowledge about the types and composition of wastes at a
disposal facility are further complications. Consequently, it
is not presently possible to use soil and waste characteristics
to predict pollutant movement/retention in soils at most
hazardous waste disposal facilities with sufficient precision
and accuracy to include the action of the soil or earth material
as a reliable part of the design for the facility. As a result,
facility location relative to groundwater sources and the
expected dilution of leachate releases in groundwater are
currently considered more important in protection of public
health and the environment than retention of pollutants by
soils. To enhance designs and predictive capabilities, future
facilities over sensitive aquifers, if not prohibited, may be
restricted to a single waste or limited wastes of known
migration characteristics. The capacity of soils and earth
materials below the facility liner to retard the movement of
pollutants may be used in future (5 years) designs, but most
likely as a safety factor rather than as a part of the primary
facility design.

Current research is intended to develop tools for designing
soil liners that limit pollutant release to a predictable
minimum. The application is primarily for landfills, but the
information is also useful in design of soil liners for surface
impoundments and in predicting the amounts of pollutants that
will pass below the zone of incorporation at land treatment
sites. The work is proceeding in two parallel phases. In the
first, methods for predicting movement of liquid (leachate) are
being developed and improved and in the second, methods are
being developed for predicting how well soil liners will remove
pollutants from any leachate that seeps through the liner.

Work is in progress on empirical predictive techniques using
samples of wastes and soils from the location of interest. This
approach appears to be the most promising way to integrate the
effect of waste and soil characteristics on pollutant retention
processes. The precision and accuracy that will be achieved by
this method remains to be determined.

In FY 1981 a Technical Resource Document (TRD) on
"Evaluation of Landfill and Surface Impoundment Performance"
(SW-869) for use by permit writers was published for public
comment. This TRD is for predicting how much leachate will be
intercepted by a leachate collection system and how much
leachate will seep thru the soil liner into underlying soils.
Following analysis of public comment, this TRD will be finalized
in FY 1982. A related TRD on methods for measuring flow
properties (permeability and diffusivity) was submitted for peer
technical review in FY 1981; it will be issued for public

comment in FY 1982 and finalized in early FY 1983. Research
will be continued through FY 1985 to test the procedures in
these two TRDs against field experience from the permit program
for RCRA Section 3004. They will be updated as necessary.

At least two additional TRDs on using batch and soil column
techniques to predict pollutant removal from leachates by soil
liners will be issued for peer technical review in FY 1982 and
for public comment and final publication in FY 1983. Field
verification of this information will be achieved by a large
scale field project and by analysis of data collected under the
hazardous waste permit program.

Pollutant Control

The overall objective of this research activity is to reduce
the impact of pollution from waste disposal sites by technology
that minimizes, contains, or eliminates pollutant release and
leaching from waste disposed to the land. The studies are
determining how various soil, synthetic and admixed materials
may be utilized as liners to contain and prevent leachates from
migrating from landfill sites.

Procedures for selecting, placement and predicting the life
of liners in landfills and surface impoundments have been the
main thrust of research efforts prior to FY 1981. Information
of these procedures is contained in the TRD "Lining of Landfills
and Surface Impoundments" (SW-870). In FY 1982 this TRD will be
revised based on public comments and updated as additional
research data become available.

In FY 1982 the research program will undertake studies of
advanced liner designs, including emerging synthetic materials
and the use of nonpetroleum-based materials (such as sulfur/
hydrocarbon compounds) in place of asphalt. Studies will be
initiated to evaluate the seaming techniques necessary to
adequately seam a variety of liner materials under various
environmental conditions and the quality control necessary to
insure the integrity of the seams. Studies will continue to
ascertain the proper placement procedures and the use of
non-soil materials to assist in bedding requirements for
synthetic membranes and leachate and gas collectors. Studies
will be initiated to develop a non-destructive leak evaluation
technique of in-situ materials. Repairs of damaged/leaking
liners will be evaluated from a quick patch, retrofitting or
complete replacement basis.

Another major effort in FY 1982 will be the identification
of agencies actively involved in liner technology research for
the purpose of data interpretation on a national scale. This

information will be utilized to standardize tests and materials
for industry use.

The research program will shift from laboratory and pilot
scale operations toward field verification studies in late FY
1982 and early FY 1983. Field conditions and their effects on
liner materials will identify problems either as improper liner
selection or placement and mismanagement of liner systems. The
field surveys will also identify why certain systems continue to
perform as based upon original design concepts. As this data
becomes available the TRD will be updated. It is also
envisioned that additional TRD's could be generated on specific
liner technology achievements.

Pollutant Treatment

The objective of pollutant treatment work is to identify the
quality of leachates being generated from hazardous waste
landfills and to develop treatment schemes for handling the
collected leachate.

The use of a leachate collection and removal system to
minimize the seepage of leachate into underlying soils is an
integral part of the landfill concept for hazardous waste
disposal. Consequently, research has been initiated to develop
procedures for handling and treating collected leachate to
insure that it is disposed of safely.

Leachate treatment schemes will likely include biological,
physical, and chemical processes; but the combination of
processes and specific modification will be greatly influenced
by the characteristics of the leachate produced. Similarities
will exist between the combination of processes selected for
leachates from co-disposal and leachates from strictly municipal
refuse. The leachate characteristics from co-disposal
facilities may be expected to be very similar to municipal
refuse leachate. Consequently, completed studies of MSW should
be pertinent and useful in selecting the appropriate treatment
schemes.

In FY 1981 a TRD "Management of Hazardous Waste Leachate"
(SW-871) was issued for public comment. Also, pilot scale
treatment studies with several waste leachates and contaminated
groundwaters were initiated.

During FY 1982 the TRD will be revised to incorporate public
comment and interim results from treatment projects. During FY
1982 and 1983, further studies to develop physical, chemical,
and biological approaches to treatment of leachates will be
initiated. The research will focus on protocols for matching

leachate quality with the appropriate combination of treatment
elements or unit processes. During this period additional field
studies will be initiated to validate results obtained from
bench and pilot scale studies. By FY 1985 the TRD will be
updated to take into account results from the various research
programs and experience with use of the TRD in the hazardous
waste permit program.

Waste Modification

Chemical stabilization is achieved by incorporating the
solid and liquid phases of a waste into a relatively inert
matrix which exhibits increased physical strength and protects
the components of the waste from dissolution by rainfall or by
soil water. If this slows the rate of release of pollutants
from the waste sufficiently and no serious stresses are exerted
on the environment around the disposal site, then the wastes
have been rendered essentially harmless and technical
restrictions on siting will be minimal.

The objective of the waste modification work is to validate
the performance of chemically stabilized wastes. This
validation is addressing four basic goals:
(1) Improvement in handling and physical characteristics of
waste materials.
(2) Decrease in the surface area across which transfer or
loss of contained pollutants can occur.
(3) Limit solubility of any pollutant contained in the
waste.
(4) Detoxify contained pollutants.

One of the options being investigated is alteration of the waste
to minimize or prevent release of pollutants from the waste into
the landfill. Fixation converts the waste to a physically
stronger and/or less soluble form while encapsulation
incorporates the waste into a strong, low solubility matrix.
Encapsulation, the more durable and expensive process, is
designed for low-volume, high-toxicity wastes.

In FY 1981 a TRD entitled, "Guide to the Disposal of
Chemically Stabilized and Solidified Waste" (SW-872) was
published for public comment. Although it contains some
information on encapsulation, it deals mainly with fixation
processes and gives procedures for evaluating the suitability of
combinations of wastes and fixation processes. A field survey
of 7 to 10-year old sites containing chemically fixed wastes was
initiated to develop verification information for the TRD.

In FY 1982, information will be developed on fixation
processes for organic wastes and work initiated on development

of procedures for evaluating fixation process suitability under
long-term (100 years) worst-case scenarios. Studies of
encapsulated waste products and production of state-of-the-art
for encapsulation processes will be completed. In FY 1983, the
TRD dealing with chemical stabilization (SW-872) will be updated
to include public comment and research results.

LAND TREATMENT

The land treatment research program has three basic
objectives. These are a) to determine environmental effects
from hazardous waste land treatment facilities and their
monitoring requirements, b) to determine hazardous waste land
treatment facility closure requirements and post-closure
monitoring requirements, and c) to develop basic knowledge
required to fully understand land treatment of hazardous waste.

Environmental Performance

Research is being conducted to determine the effects of
hazardous waste land treatment sites on groundwater, surface
waters, soil/plant systems, and air emissions. Measurements at
existing and closed sites will help evaluate the environmental
performance. Because the petroleum industry has a longer
history of land treatment usage, cooperative projects with the
industry are being conducted to provide short-term useful
information. Data from closed petroleum land treatment sites
will provide a more complete picture of post-closing
environmental effects, help to better define active site
monitoring requirements, and help to evaluate the data from
active sites. Measurements at operating sites will help to
determine actual environmental effects existing today and how
they are being produced. They will also provide limited
opportunity for evaluating monitoring techniques, e.g., soil
pore water, soil monitoring, air monitoring, etc., and
additional monitoring development needs such as the proper
placement and use of lysimeters. More controlled research at
field sites is needed. Therefore, several sites will be
selected and monitored from initiation to completion in order to
obtain controlled data for the total life cycle of a land
treatment site.

Closure/Post-Closure

Improperly closed sites may create conditions that cause
adverse environmental effects for years. Yet, imposition of
unrealistic closure procedures may result in loss of potential
hazardous waste disposal and/or treatment facilities because of
economic hardships. Therefore, the research includes

determining site closure procedures required to assure
environmental protection from potential pollutant movement.
Data generated in measuring the environmental effects of closed,
existing, and future sites will have important input to meeting
this objective. Also important is determining what facility
monitoring is needed after post-closure to ensure proper
environmental protection. The research includes critical
evaluation of existing data and longer term research to
determine types and characteristics of hazardous waste
applicable to land treatment. Also important is the
identification of the long-term pollutant potential of land
treatment, including pollutant movement and control.

Research will provide data applicable to site closure and
will evaluate and compare four closure techniques: (1)
revegetating the site, (2) closing the site as if it were a
landfill, (i.e., covering), (3) removing the pollutant
contaminated soil, and (4) leaving the site as is. Measurements
will be made to determine the effectiveness of each closure
procedure to prevent or control the movement of remaining
pollutants from the closed site. In the case of covering as a
landfill, measurements will be required to determine when cover
placement should take place to assure that the cover does not
stop further degradation of organic pollutants. Verification
will be sought on the effectiveness of vegetation to control
pollutant movement from the site by runoff, air erosion, and
associated mechanisms. These efforts will also provide data on
the relative economics associated with the different techniques.

Coring samples at active and closed sites (if available)
will be needed to provide data to help determine if downward
movement of pollutants (primarily organics) is a potential
problem at those sites. Data generated will also help determine
actual degradation rates of some of the pollutants. Some
cursory data has indicated that selected organic contaminants
(primarily from the oily waste sites) have not degraded as
rapidly as originally thought. More study will be required to
verify such findings.

Waste, Soil, and Plant Interactions

Available data for understanding hazardous waste treatment
interactions are very limited. Longer term research is needed
that will build the data base necessary to determine the proper
role of land treatment for managing hazardous waste.
Laboratory, greenhouse, and field studies with specific chemical
waste will help determine effects of pollutants on food and
nonfood chain crops; development of soil/crop response curves;
determination of hazardous pollutant movement; ion exchange
capacity effects; soil absorption effects; soil neutralization

effects; waste preconditioning; and other factors requiring
study to better understand land treatment of hazardous waste.
These results will also help characterize industrial sludges
that are land spread, particularly with respect to the organic
constituents. This will help identify those wastes requiring
special management. Interactions of wastes with soils and
plants will be done by examining factors governing competition
of clays and organic materials in the wastes;
absorption-desorption and movement of selected wastes in soils;
degradation and persistence in soils of selected wastes;
phytotoxicity of selected elements on selected crops; waste and
soil pretreatment mechanisms, assessment of bioassay techniques,
and others. These studies will help optimize and expand the
current technology, primarily used by the petroleum industry, to
other industrial wastes (i.e. the pharmaceutical industry,
emerging energy production technologies, organic chemicals, and
others).

The research program to date has developed a technical
resource document for "Hazardous Waste Land Treatment"
(SW-874). The document is based upon existing information and
discusses hazardous waste land treatment site selection and
design requirements, soil characteristics, waste
characteristics, application rates and techniques, operating
procedures, monitoring requirements, closure procedures, and
other factors important in the proper operation of hazardous
waste land treatment sites. As new data become available, the
document will be revised.

SURFACE IMPOUNDMENTS

The surface impoundment research program will be modified to
provide a comprehensive understanding of the design, operation,
and maintenance of surface impoundments as options for hazardous
waste disposal. Information is needed on the use of natural
soils as liners and the correlation of laboratory measurements
with the construction standards achieveable in the field. Of
particular interest is the degree to which specification of
construction techniques and inspection practice can influence
uniformity and performance of the finished impoundment.

Containment

The use of natural soils as liners is attractive because
on-site availability can lower construction costs. However,
waste-soil interactions can pose problems in addition to those
normally involved in earthen structures. The potential contact
of soluble/liquid organic compounds with soil is much greater in

surface impoundments than in other disposal facilities. Any resulting changes in soil structure or liquid permeability are serious because of the large amounts of liquid present and the substantial hydraulic head. Work is underway on predicting the effects of organic wastes on soil properties. This will be continued and opportunities for field verification developed. The emphasis is on production of testing methods rather than on comprehensive data concerning a variety of waste and soils.

Work on the movement of pollutants through soil liners is being conducted as part of the work on landfills; work on the effects of organic wastes on the durability and other properties of soil and synthetic lining materials is also included in that work area. The information developed will be applied to surface impoundments as well. Procedures for evaluating the effects of wastes on natural and synthetic lining materials and earthen structures that are unique to surface impoundment, for example wave action, aeration and mixing, sunlight, and deposition in the impoundment will be evaluated.

Because of the large ratio of exposed surface area to volume in surface impoundments, the potential for emission of volatile organic compounds into the ambient air is of particular concern. A procedure for predicting air emissions as a function of waste composition, impoundment design, and weather conditions is being developed. A field monitoring program will verify the predictive method. A final phase in this work will be the development of a TRD addressing the use of the predictive model and the field verification procedure.

Another area of concern is the siting/size problem. The socioeconomic aspects of this problem will be addressed in tandem with similar research effects for landfills and land treatment facilities. A complicating constraint that must be addressed is the economic disincentive to locate a surface impoundment at any great distance from the source generating the waste. The size problem is independent of location and requires development of methods for assessing the tradeoff between increased potential for environmental damage (area seepage, dike failure) from larger impoundments and the economies of scale achievable as impoundment size is increased. Initial studies will examine the magnitude of these factors for a series of smaller impoundments containing the same total volume as a single large impoundment.

Closure of surface impoundments has been an initial concern because of the uncontrolled dump site problem. As such, a TRD "Closure of Hazardous Wate Surface Impoudments" (SW-873) was published in FY 1981 for public comment and review.

Performance Evaluation

Additional Technical Resource Documents (TRD) are being developed for conducting performance evaluations of surface impoundment under different designs. The source of information and data will be generated from field surveys where the advantages and disadvantages of specific designs, management procedures, geographic location and waste contained are identified and evaluated. This data set is envisioned to produce information that can be modeled to predict the success or failure of future designs. This is expected to be available in early FY 1984.

ECONOMIC ASSESSMENT

The objective of the economic assessment work is to provide realistic cost choices for user communities for selection of various waste disposal options.

The economics of disposal technologies have always been of concern for each technique. With the development of several options for disposal, the cost benefits/comparisons need to be assessed and updated to more realistically provide optimal choices for disposal. As new scenarios are developed, the related costs in comparison to established procedures need to be understood.

Risk/benefit analyses are being developed to provide standardized procedures associated with the disposal of hazardous wastes. Technologies will be rated and compared in terms of susceptibility to catastrophic events, unexpected downtime and adverse environmental impacts. Other risks associated with the existence and operation of each technology will also be assessed. An incentive/disincentive program for disposal technologies will be evaluated for overall impact.

Second and third generation disposal options and the part they may play in implementating future regulations will be assessed. Consideration of the technical and institutional changes, combined with a more exact analysis of the disposal economics, will provide a solid foundation for the economic impact analysis required in conjunction with new legislative and executively mandated thrusts.

TECHNICAL RESOURCE DOCUMENTS

The Technical Resource Documents (TRDs) which have been

mentioned briefly in this paper consist of a series of documents which represent Best Engineering Judgement (BEJ) for the design, operation and closure of hazardous waste disposal facilities. These documents are a compilation of research efforts to date and are being developed to assist in the implementation of 40 CFR, Parts 264, 265, and 267 regulations concerning hazardous waste disposal facilities (landfills, surface impoundments, and land treatment). These documents are technically oriented and not policy oriented. They are not regulatory by design, but are intended only to provide guidance in decision making processes. The documents will be updated continuously as the bases for understanding and need changes. Some of the manuals provide specific design and operational information (i.e. closure and lining TRD) while others are intended as evaluation tools (i.e. hydrologic simulation and performance evaluation). Eight TRDs have been completed to date. Six of these documents are related to landfills and one document each is related to surface impoundments and land treatment. A listing of these eight documents by title and report number along with a brief description of document content is listed below:

TRD 1, Evaluating Cover Systems for Solid and Hazardous Waste (SW-867): This document presents a procedure for evaluating cover systems for solid and hazardous waste. It contains a checklist, with supporting documentation, that a permit writer can use to evaluate a proposed cover system.

TRD 2, Hydrologic Simulation on Solid Waste Disposal Sites (SW-868): This document provides a computer package to aid planners and designers by simulating hydrologic characteristics of landfill covers to predict percolation as a function or cover design and climate.

TRD 3, Landfill and Surface Impoundment Performance Evaluation (SW-869): This document describes how to evaluate leachate collection systems using compacted clay or synthetic liners to determine how much leachate will be collected and how much will seep through the liner into underlying soils. The adequacy of sand and gravel drain layers, slope, and pipe spacing are also covered.

TRD 4, Lining of Waste Impoundment and Disposal Facilities (SW-870): This document provides information and guidance on liner systems. It discusses waste, liner types, compatibility, liner slelection, specifications, design of leachate collection systems, and case study analysis methodology. It also includes a glossary of liner system related terms.

TRD 5, Management of Hazardous Waste Leachate (SW-871): This document discusses leachate composition, leachate

generation, selected management options, available treatment
technologies, unit-treatment options, and system economics. It
presents management options that a permit writer or hazardous
waste landfill operator may consider in controlling and treating
leachate.

TRD 6 Guide to the Disposal of Chemically Stabilized and
Solidified Wastes (SW-872): This document provides basic
information on stabilization/solidification of industrial waste
examines the regulatory considerations, current and other
options involved in disposal systems using stabilization/
solidification of wastes. A summary of the major physical and
chemical properties of treated wastes is presented. A listing
of major suppliers of stabilization/solidification technology
and a summary of each process is included.

TRD 7, Closure of Hazardous Waste Surface Impoundments
(SW-873): This document discusses and references the methods,
tests, and procedures involved in closing a surface impoundment
in a manner that (a) minimizes the need for further maintenance,
and (b) controls, minimizes, or eliminates, to the extent
necessary to protect human health and the environment,
post-closure escape of hazardous waste, hazardous waste
constituents, leachate, contaminated rainfall, or waste
decomposition products to groundwater, surface waters, or the
atmosphere. Problems that have been overlooked in abandoned
impoundments and have caused environmental degradation are
discussed. The document discusses closing an impoundment either
by removing the hazardous wastes or by consolidating the waste
onsite and securing the site as a landfill. Relevant literature
or procedures are documented for more in-depth review as
necessary.

TRD 8, Hazardous Waste Land Treatment (SW-874): This
document discusses land treatment medium, characteristics of
hazardous waste, allowable waste loadings, facility planning for
land treatment, land treatment permit process, closure, and
items to be considerered during permit evaluation.

CONCLUSIONS

This paper has presented an overview of the hazardous waste
land disposal research program as conducted by the USEPA.
Specific areas discussed in order of priority were landfills,
land treatment, and surface impoundments. All activities in
these areas are being summarized into technical resource
documents to provide guidance in decision making processes.
These technical resource documents will be routinely updated to
reflect the latest research findings and it is planned to

condense these documents into guidance manuals. A guidance
manual will be prepared for each subject: landfills, land
treatment, and surface impoundments. More information about
specific aspects of the hazardous waste land disposal research
program can be obtained by contacting the author.

HAZARDOUS WASTE LANDFILL PRACTICE

K.A. Childs

Environmental Protection Service
Ottawa, Ontario, Canada

BACKGROUND

It is difficult to identify when the topic of Hazardous Waste Management assumed a position of national and international interest and concern. What can be recorded is that in the early 70's several countries realized that many abandoned and currently operating disposal sites were the cause of, or had the potential for causing, the contamination or pollution of vital ground and surface water resources. In view of the fact that landfills were identified as one of the primary sources, or potential sources, of pollution, it was concluded that to expect landfills designed to accept mixed municipal solid wastes to also accept the increasing quantities of dangerous and toxic waste materials being brought to them was not entirely a responsible position. Severe environmental problems were a real possibility. Recognition of the need to better understand all the facets of land disposal lead to positive action by various countries and created an environment conducive to international co-operation.

It was the Federal Republic of Germany which proposed, in 1973, to initiate pilot studies on the management of hazardous wastes and further proposed that the task be entrusted to several industrialized countries who were, and remain, members of NATO. This position was approved by the NATO Council in November 1973, and the work was actually performed in two phases as follows:

Phase I (1973 - 1977)

- Landfill research and landfill practice
- Underground disposal
- Transportation
- Organization
- Recommended procedures for hazardous waste management
- Chromium cycle

Phase II (1978 - 1981)

- Thermal treatment
- Chemical, physical and biological treatment
- Landfill research (continuation of Phase I)
- Metal finishing wastes

The workload was divided amongst the various participating countries. Canada was asked to act as a "lead country" for the "landfill research" topic for Phase II, representing a continuation of the work initiated by the United Kingdom under Phase I.

You will note that the title of my presentation is "Hazardous Waste Landfill Practice". Landfill practice formed only part of one of the tasks in Phase I with the bulk of the effort in both Phases I and II being devoted to landfill research. The objective of my presentation will be to summarize the landfill practice aspects of the various studies and relate certain aspects of the research program to the practice of landfilling.

EXTENT AND IMPORTANCE OF LANDFILLING

The extent to which landfilling and special chemical landfilling of hazardous wastes is practised varies from one participating country to the other and even within the borders of some countries. Generally, at present, it represents an important method of disposal in almost all countries, and this situation is likely to continue for the foreseeable future.

The following summary is provided to indicate the extent that landfilling is used a management tool in the various participating countries.

In the absence of any real alternative disposal methods having similar high capacity and ability to cope with large volumes of a very wide range of wastes, it was concluded that the importance of landfilling is not likely to be lessened. As new technologies evolved and risks became identified limitations could be imposed on landfilling as a disposal means particularly for the more hazardous wastes.

- Landfilling in Denmark is generally limited to being used as a disposal method for municipal and non-hazardous industrial wastes. Hazardous and toxic wastes are treated by methods with such as physical/chemical, incineration with the residues of these processes being landfilled.

- France considers landfilling as an acceptable means of disposal preferably when the nature of the waste is fully understood. This implies that work is required in determining the nature of the waste and the interaction with soil once it is placed in the ground.

- The Federal Republic of Germany considers landfilling to be an important part of all waste management systems. 5 categories of landfills are identified.

(a) Municipal landfills - generally limited to receiving municipal wastes.

(b) Multi-component landfills - authorized to receive municipal wastes plus compatible industrial wastes and hazardous wastes under special conditions.

(c) General industrial landfills - limited to receiving industrial wastes and generally not used for municipal solid waste.

(d) Mono-component landfills - used when separation of wastes is desirable to allow possible future recovery.

(e) Special wastes facilities - where the landfill would be used almost solely for receiving residues for incineration, chemical treatment, etc...

- In the Netherlands landfilling is permitted in the presence of strict requirements to protect the ground water. The preferred means of disposal are regeneration and re-use, treatment and incineration, special storage including the use of abandoned salt-mines. Landfilling of hazardous wastes is only allowed in the presence of strict special conditions. It is preferred that landfilling be performed above ground level rather than in excavations or filling in of depressions. This preferred method appears to be a fundamental management tool - for both protective and operational reasons.

In Norway, despite topographic and climatic difficulties, landfilling is a significant waste management tool. However, by design, only 5% of special wastes are directed to municipal landfills. The balance of the special wastes is directed to regional landfills dedicated to receiving these special wastes.

- 90% of all "controlled wastes" in England are deposited at landfills. Clearly disposal by landfilling is the more important component in all waste management systems in England. Only 1-2% of these controlled wastes are deemed to be hazardous. Principal control of waste management in England is exercised through legislation and "advisories" passed or issued, by the senior (Federal) government.

The U.S.A. enjoys a similar range of topographic, climatic and demographic conditions as does Canada, and, similarly landfilling represents a major element in nearly all waste management programs.

Federal control over waste management and hazardous waste management is exercised through RCRA which requires that each individual state file, with EPA, management plans which must meet the federal EPA standards.

- In Canada, at present, it is the principal method of disposal used by almost every municipality with only a few municipalities having any other alternative means for managing even selected fractions of the waste stream. Facilities for the disposal of wastes which might be considered special or hazardous are limited and are becoming increasingly more difficult to locate or construct due to adverse public reaction.

- Control over waste management in Canada comes within the purview of the provincial governments with the federal government being involved only when there is interprovincial or international movement of waste. The Canadian government's efforts in waste management are devoted principally to factors such as research and technology development and demonstration, inventories, problem identification, and assisting in controlling cross boundary shipments of certain wastes.

Based on the preceding brief overview it appears that in the NATO countries two approaches have been adopted:

(a) Landfill is almost universally adopted as one of a number of acceptable options for waste disposal, and it generally represents a major element in the overall disposal strategy for various wastes including domestic, commercial, industrial and certain types of hazardous wastes.

(b) In the absence of a viable alternate, in jurisdictions where
     restrictions on landfilling of chemical or hazardous wastes
     may apply, either due to legislation or due to the limited
     number of suitable available sites, exceptions to the
     limitations may be allowed in special cases.

     The question of options are more fully discussed in some
detail in the report of the sub-project on "Recommended Procedures
for Hazardous Waste Management", carried out as part of the pilot
study on disposal of hazardous wastes.

     Continued use of landfills for the disposal of certain
hazardous wastes and special chemical landfilling to varying
extents seems to be a generally accepted principle in all
participating countries because of the considerable geological,
hydrogeological, topographical, climatological and population
density differences between the participating countries it may not
be possible to devise a single "Management" approach to landfill
of hazardous wastes. It appears that most countries will seek to
optimise their use of landfill starting from one or other of the
basic approaches mentioned above.

IMPORTANCE OF WATER RESOURCES

     The importance of avoiding harmful effects of landfills was
recognised by all participating countries. The most important of
the possible adverse effects was identified as the potential
pollution of water resources, and in particular the contamination
of groundwater. Relevant parameters with respect to water
pollution include the properties of the wastes, such as
solubility, acidity, any action as a solvent, and chemical
reactivity giving rise to different substances; also the capacity
for absorption, adsorption, biological or chemical breakdown and
ion-exchange processes which may exist in the landfill itself or
in other materials encountered en route to any relevant waters;
finally the use to which any such waters are put, and the dilution
factors which they provide.

     To indicate the variation in relative immediate importance of
ground and surface water resources, it was noted that in Canada
potable water supplies were obtained from surface waters for 90%
of the population and from ground waters for the remaining 10%.

     This compares with Denmark where almost the exact inverse is
the case, only 2% of the Danish population is served by water
supplies derived from surface waters - 98% is derived from ground
waters.

France reported that 52% of public water supplies come from surface waters and 48% from ground waters.

Regardless of current differences in the rates of utilization of ground water as a source of potable water supplies it was concluded that protection against contamination of ground and surface waters was of paramount importance.

Participating countries cited incidents where ground and surface waters had been polluted due to landfilling activities. Examples included the following:

(a) The United Kingdom reported on the use of a disused granite quarry for the disposal of food wastes and the residue from plasterboard manufacturing. When the mix became saturated a severe leachate problem was created due to anaerobic decomposition of the combined wastes. Remedial actions implemented included the provision of systems to aerate and chemically treat the leachate prior to discharge to the sewerage system. Also, in England, the use of an area for landfilling where the geological environment comprised of fissured limestone and soft calcereous sandstone which represented an important aquifer, resulted in the creation of a leachate plume. Fortunately the limit of the plume's migration appears to have been determined by natural processes within the soils.

(b) In report number 64, the United States identified 4 incidents of pollution caused by the land disposal of hazardous wastes including an instance where a shallow water table was contaminated by the by-products from the manufacture of pesticides, herbicides and chemical warfare agents.

An increase in the heavy metal concentration in ground water was attributed to allowing the use of an unlined disused sand pit as a disposal site for acidic and oily wastes. In addition to this there was a failure of a dike allowing contamination of a major waterway. Remedial action has been taken.

A major aquifer in New Jersey was contaminated when storage drums containing petro-chemical wastes leaked allowing the migration of noxious liquids into the soils. Another recorded incident involved the migration of arsenic wastes from a site. Contamination of potable water supplies was limited due, principally, to the natural upward hydrostatic pressure in the vicinity of the site. As an added precaution, the use of the aquifer within a 15 miles radius of the site was restricted for up to 10 years.

Since 1977, a significant number of problems have been identified in the United States including the Love Canal, the Hyde Park site in the Niagara area and the Valley of the Drums. In addition to these now famous cases, many potentially hazardous sites have been located and remedial action will be undertaken using the recently established superfund.

(c) In Germany, the use of a disused gravel pit as a landfill site was followed by the migration of cyanides wastes which contaminated a water supply. This resulted in the need to treat the water before it could be used for domestic purposes. To further alleviate the situation it was also decided to remove the waste from the site.

Another case involved also cyanide wastes. Wastes were deposited without authorization causing pollution of surface waters. Fortunately, due to the fact that the site was located in an area where the soils were relatively impermeable, ground waters were not seriously affected.

There are also recorded cases of interactions between a leachate and the native soils resulting in a significant change in the quality of the ground water.

When such incidents have been experienced, they have been due to factors such as poor site selection, inadequate supervision, and incorrect methods of deposition as well as insufficient knowledge about soil-waste interaction in the underlying strata, rather than to basic unsuitability of landfill as a disposal method. With the advent of new legislation providing for the control of hazardous waste in many participating countries it is anticipated that better hazardous waste landfill practices will be employed in the future.

It is anticipated that the situation of identifying and correcting the errors of the past will be a continuing and universal challenge.

ATMOSPHERIC POLLUTION

In addition to those problems created by the migration of, or contact with, contaminants adversely impacting on water resources, another hazard identified by a number of countries was the release of possibly volatile vapours into the atmosphere. This risk was considered to be short term but further research work was recommended.

SITE SELECTION CRITERIA AND SITE SUITABILITY

Site selection and preparation was addressed in reports
numbers 62 and 64. In report number 62, there is extensive
discussion of the site selection process and some of the
socio-political mechanisms.

Tabulated in report #62 are 88 criteria for site selection.
These criteria, in various combinations, were being taken into
consideration.

14 of the 88 criteria were commonly identified by the
majority of the agencies. These 14 might therefore be considered
as the most important. Predictably, amongst those given highest
priority were those criteria which relate to the protection of
water quality. These 14 criteria are as follows:

- Effect on drinking and industrial water supply.
- Effect on wastewaters and percolating water drainage.
- Effect on surface water drainage.
- Imperviousness of subsoil (permeability)
- Water resources in surrounding area.
- Surface water drainage.
- Flood protection.
- Expected quantity of percolating water.
- Suitability of soil.
- Landslide, earthquake hazards.
- Effect on long-term surface water drainage.
- Depth to water table.
- Ion exchange capacity of soil.

The 88 criteria referred to earlier were all developed in an
attempt to address the following factors:

1.   Protection of the environment particularly protection against
     contamination of water resources by leachate.

2.   The suitability of the site for receiving particular types of
     wastes.

3.   The ultimate use of the land after closure.

As a general statement it appears that two philosophies have
influenced site selection as it relates to the ability of the site
to tolerate wastes:

(a) To select sites on the basis of the site's ability to contain wastes and leachates, especially leachates from hazardous materials.

(b) To select sites with the knowledge that containment within the site could not be achieved but that natural processes both within the site and outside the site would attenuate and/or disperse contaminants and reduce contaminant concentration in any leachate.

Sites relying on containment are typically those located on impermeable strata such as clays, but the ability of sites located on other types of strata to contain wastes and leachates can also be enhanced by the use of site liners. Collection and treatment of leachate must be accepted as a necessary part of a containment scheme. For disposal of particularly hazardous types of waste in sensitive geological or hydrogeological areas, containment may be essential. For example, the storage of radioactive waste will require total containment. Use of such measures may in other situations be unnecessary due either to having precise knowledge of the soil waste reaction and the behavior of the migrating contaminants.

Use of the sites relying on containment for disposal of significant quantities of liquid wastes, or sites in areas of high natural precipitation, may cause problems in operation (such as difficulty in achieving a suitable liquid to solid ratio) and result in the build-up of a large volume of contaminated leachate within the site; stabilisation or restoration of the site after completion of landfill operations may involve extensive and costly remedial measures.

In addition to what might be referred to as technical and scientific criteria, other constraints which must be considered include:

1. Legislation at all levels of government.

2. Site operating conditions either specified or as a consequence of location.

3. Technical and economic factors. That is the relationship between generators, site owners, disposal authorities, etc...

4. The socio-political factors, particularly public acceptance.

The question of public acceptance was dealt with at length and included suggestions with respect to the mechanisms and techniques which might be employed in the pursuit of gaining public acceptance.

It is now universaly accepted that the public must be consulted at an early stage of the site selection process and on a continuing basis during the planning and development stages.

Probably the most important development in this regard is the recognition of different segments of society - different "publics" and their varying needs and the consequent need for varying techniques. It is also universally realized that despite enormous, well planned, and costly efforts to inform and placate the public, acceptance still remains the most elusive goal. From the time of commencing Phase I, it is reported that what was originally a North America phenomena has spread to Europe, and facilities which have existed for many years probably would not have been built if the public attitude which prevails now had prevaled when they were originally commissioned.

Also since the time of concluding the Phase I report, significant work has been done by a number of agencies in the development of programs and techniques directed towards achieving productive public involvement in the site selection process.

SITE DEVELOPMENT TECHNIQUES - PRACTICAL APPLICATIONS

There have been many attempts to develop a practical, universally applicable site evaluation program. The efforts have not been totally successful due to the fact that neither wastes nor sites lend themselves readily to classification schemes. For practical purposes, a more generalised scheme capable of modification and adaptation to meet local requirements is usually needed.

In the Phase I, three site selection techniques, or schemes, of varying complexity were described:

(a) The soil wastes interaction Matrix procedure developed by Canada:

This program was based on the development of 2 Matrices:
1. A site independent Matrix
2. A site specific Matrix.

This program would allow the user to determine the suitability of a site based on the predictable results of the interaction between the soil environment and a particular waste. For a number of reasons this technique has not received broad acceptance in Canada.

(b) Based on observations of the effect of depositing waste at selected sites, the Federal Republic of Germany identified 11 criteria for assessing the suitability of a site. The criteria addressed environmental operational and developmental factors to account for the following activities and concerns:

- Co-disposal
- Protection of waster resources
- Climatic conditions
- Gas migration and leachate control
- Soil permeability

(c) To assist in the determination of site suitability relative to the inter-action between soil conditions and the type of waste, the United Kingdom developed 3 classes of sites based on the geology of the site and the capability of the natural environment to accept wastes. These classes are:

- Sites providing a significant element of containment for wastes and leachates.

- Sites allowing slow leachate migration but providing significant attenuation.

- Sites allowing rapid leachate migration and providing insignificant attenuation.

The classification within which the site falls determines the .nature of the waste which can be disposed of at the site. It also determines to some extent the nature of site development, operational factors and after-use utilization.

SITE LINERS TO IMPROVE SITE SUITABILITY

The use of synthetic liners to make a site either suitable or as an insurance factor was addressed in the Phase I report and was referred to as "a comparatively recent development". The use and knowledge of liner behavior has changed significantly since that time. Attention is drawn to the United States EPA publication entitled "Lining of Waste Impoundment and Disposal Facilities" which probably represents the most authoritative document on the topic.

In 1977 two techniques were identified:

- Treatment of the base and walls of sites with an absorbent or impermeable substance;

- Installation of an impermeable membrane, usually made of a suitable synthetic material.

The following factors were identified as being significant and should be considered if the use of a liner is being contemplated:

- The compatibility between liner materials and the wastes.
- The physical caracteristics of the liner materials in terms of flexibility, weather and site condition, etc...
- The degree of containment required and the anticipated or required lifetime of the liner materials.
- The fact that containment might increase the need for treating leachate which cannot now migrate from the site.

GAS GENERATION AS A FACTOR IN SITE SUITABILITY

The site selection process should recognize the fact of gas production as a result of landfill operations.

3 broad considerations should be reviewed:

- The possibility of causing a hazardous situation by the migration of the gas which might collect in pockets creating the potential for an explosion.
- The potential for collecting the gas and using it as an energy source.
- The possibility of damage to vegetation and the impact on land reclamation schemes either by affecting the root growth or requiring that venting be provided.

With respect to gas generation as a predictable phenomena knowledge of the process and the capability to accomodate it has improved significantly since 1977. At this juncture it is necessary to only briefly review the current status as it applies to those 3 aspects identified earlier:

1. With respect to hazardous situations, the expertise and the
    mechanical capabilities now exist to either avoid creating,
    or to remedy, situations which are, or might be, hazardous
    due to the generation of methane.

2. Regarding collection and use, considerable work has been done
    in this area of interest prompted by the increasing cost of
    energy. In North America the most significant gas recovery
    systems have been installed in California where landfill gas
    from landfills that have accepted predominantly municipal
    solid waste is prepared to the point where it can be
    pipe-lined into conventional gas transmission systems.

    Canada has also undertaken experimental works to develop
    techniques whereby smaller landfills might be utilized
    productively for methane gas recovery.

3. The problems of causing damage to existing vegetation or the
    inability to initiate and then to sustain its growth on or
    around landfills continues to be investigated. Advances made
    in the past few years to solve problems of vegetation damage
    suggest that these problems can be minimized by proper site
    preparation and judicious selection of the type of
    vegetation.

    With respect to land reclamation, it is now commonplace for
    engineers to be called upon to either design a new landfill
    or modify an existing landfill to allow for optimum final
    land use.

4. To the commonly acknowledged problems of methane
    generation and migration must be added the problem
    of the other gases which might be released to the
    atmosphere. This can occur during the disposal oper-
    ation or at the time of excavating as part of
    corrective or remedial activities.

SITE OPERATING FACTORS

Leachate

    The leachate generating potential of landfill sites depends
on natural precipitation, the groundwater level, the properties of
the waste itself including hazardous waste, and the method of its
deposition in the landfill. Several projects have been commenced
to investigate these inter-relationships. Preliminary findings
indicated that the quantity of leachate could be reduced by

subjecting each layer of waste to a high degree of compaction and by applying layers of impermeable material (eg clay) at various stages within the landfill.

Methods for collecting leachate, treatment to render leachate suitable for discharge, and the effects of recirculating leachate through the landfill were also studied. The latter technique may reduce the quantity of leachate. The composition of leachate may also be modified and its polluting potential altered.

Factors controlling the movement of leachates away from sites are beginning to be understood. These processes involve complex interactions between chemical, physical, biological and geological factors. The effectiveness of these interactions in attenuating contaminants provides the basis of the "attenuate and disperse" principle. The use of the landfill and the land itself in controlled chemical reactions also were, and remain, an important area of study.

In the more recent investigations, the principal objective has been to develop an understanding of the processes governing movement of leachates at existing sites. It was concluded that when further information from field observations and laboratory studies is available, it may then be possible to make predictions relevant both to the study sites and to other sites for which some basic data are available. Evidence from preliminary results of field investigations and laboratory/lysimeter research work supported the view based on experience that significant attenuating and recovery mechanisms apply, particularly when systems are not overloaded.

METHODS OF DISPOSAL

All participating countries use land disposal for a wide variety of wastes in all physical forms. The wastes may include solids, sludges, liquids and combinations of all phases. These wastes are delivered to the site either in bulk or in containers. Also, because of the nature of collection systems, chemical and some hazardous materials might be included within the wastes collected as part of the regular collection service.

The fact that landfill is required to accomodate such a wide spectrum of wastes prompts the posing of two important questions:

1.   What will be the effect of depositing a particular waste at a given site?

2.   What is the best method for depositing waste of a particular
     type, i.e. in what quantities, at what rates and by which
     techniques should the waste be landfilled?

     Answers to the first question are usually based on past
experience or from results of specific research programs.

     The second question has important implications for site
operation and management.

HANDLING METHODS FOR PARTICULAR WASTES

Liquid Waste Handling

     The disposal of liquid wastes to land with proper supervision
can be a satisfactory method of disposal.

     Improper discharge to the land, either by virtue of
inadequate supervision, unplanned co-disposal or an unsatisfactory
site in the first instance, can cause serious short-term and/or
long-term problems.

     Three methods of discharging liquid wastes at landfill sites
were identified:

(a) Direct discharge into the working face.
(b) Discharge into a system of channels, trenches or distribution
    headers.
(c) Discharge of liquids into special lagoons.

     The first 2 methods depend upon the absorptive capacity of
the wastes in the landfill which in turn is determined by the
composition of the waste and the transmission of water (either
direct rainfall or surface waters) through the site. One standard
of absorptive capacity cited by the United Kingdom was 0.15 m^3
per ton of waste in place, based on an in place density of 500
kg/m^3.

     Another factor mentioned was the importance of the contact
time between the waste and the liquid. This is particularly valid
when liquids are discharged directly to the working face. It
becomes of less importance when trench, and pipe distribution
systems are employed.

     Disposal by lagooning requires considerable surface area and
may be less desirable due to the risk of odors, spillage and

leaking through the base. Lagoons do have the advantage of being able to receive volume of wastes of preferably known composition.

Frequently, however, lagoons are used as vessels for receiving wastes of different composition making final disposal or treatment more difficult. Lagoons can represent a safety hazard which can be minimized by good site management and monitoring of the wastes entering the site to ensure either proper mixing or separation. Special precautions to be employed when handling liquid wastes were identified as follows:

(a) Ideally, low flash point liquids should not be deposited at a landfill site. If these wastes are brought to the site their discharge should be strictly supervised.

Flammable liquids should always be deposited away from the active working face.

(b) The possibility that acids should be neutralized before discharge should be investigated. Co-disposal should be undertaken with care to guard against dangerous reactions.

(c) Alkaline materials such as caustic alkalis should be isolated from ammonium salts and phenolic compounds.

Sludge Management

Most sludges disposed of at landfill sites are the product of chemical precipitation processes associated with process effluents such as metal finishing sludges, tank bottoms and cell muds.

The chemical composition of sludges varies as is the case for liquid wastes. Physical composition varies from thin slurries (1-2% by weight solids) to de-watered residues (30-50% by weight solids). Sludges have potential for causing water pollution, but problems are more likely to occur due to difficulties of handling. Two methods of discharging sludges were identified:

(a) Dispersal Methods

This is when the sludge is disposed of evenly throughout the site, usually along the working face. This method is usually employed at sites where only domestic, commercial and inert wastes are received.

(b) Segregation Methods

This is used when a specific area of the site can be used to avoid accidental and, possibly dangerous, mixing with other materials resulting in the creation of a major problem.

Sludges disposed on in this manner might be subjected to chemical treatment or a fixation process to minimize either the risk of pollution or the mobility of undesirable contaminants.

Generally wastes in liquids or semi-solid form should only accepted at a site (1) when the absorptive capacicy of the receiving media is known and not exceeded, (2) when the possibility of rapid movement of the sludge through the landfills to aquifers is unlikely, (3) when reaction between various wastes can be avoided and (4) when the ultimate use of the land will not be adversally affected by the deposition of special wastes; for example potential harm to grazing animals and adverse effects on root crops.

It was noted that when there was a potential for recovery of materials of value, sludges and liquids preferably should not be landfilled.

Drummed Wastes

The use of drums as containers for wastes should be limited to containment of inert wastes. Problems caused by the eventual physical or structural failure of drums can be minimized by judiciously selecting materials which might be placed in drums.

Currently many sites are being identified where drummed wastes have been deposited in the past and identification and the management of the wastes contained in the drums is most difficult.

The implementation of remedial or corrective measures at these sites can be both hazardous and costly. The use of drums for waste containment in land disposal operations was generally not advocated.

Controlled co-disposal of wastes

In 1977 when report #64 was completed, it was noted that there was "considerable scope for developing co-disposal techniques and technologies". The situation remains the same in 1981.

It was reported that subject to suitable sites being available, co-disposal could represent an important factor in waste management strategies. Effective co-disposal at environmentally safe sites could reduce or at least modify the risk.

In addition to operational and environmental benefits, logistical benefits might be realized by reducing transportation costs. As noted earlier, co-disposal should not be used when there is a potential for recovering valuable resources from the waste stream, nor should it be used for highly noxious wastes. Noxious wastes should only be landfilled when no other system is available.

Examples of planned or controlled co-disposal include:

1. Federal Republic of Germany

(a) Disposal of metal hydroxide sludge with municipal solid wastes - sludge is deposited in earth lined pits dug into the municipal wastes. The pits are then filled with the sludge and covered with soil and additional municipal wastes.

(b) Acid tar residues co-disposed with municipal wastes - sludge is deposited in basins near the working face where the acid tar reacts slowly with the municipal wastes.

(c) Industrial liquid wastes and sludges co-disposed with municipal wastes - this is usually undertaken in clay lined lagoons excavated in the municipal wastes.

2. United Kingdom

The United Kingdom reported that the following wastes are accepted for co-disposal with municipal wastes:

(a) Metal hydroxide sludge
(b) Asbestos wastes
(c) Solvent contaminated residued
(d) Soils containing spent oxide
(e) Dry cleaning wastes
(f) Printing ink sludges

Generally it is permissible for wastes which are regarded as non-hazardous including water treatment sludges and certain industrial treatment sludges to be incorporated into domestic wastes.

One commonly accepted controlling factor is the presence of heavy metals in the sludges. This concentration should not be significantly greather than the concentration in the domestic wastes or the soil.

Currently co-disposal is practiced by both the private and public sector, generally using mature domestic wastes - i.e. wastes which have been in place for some time.

3. United States

Disposal of liquids and sludges either separately or as a component in a co-disposal operation are subject to the provisions of RCRA. It was reported that certain states have classification systems which allow for co-disposal under suitable conditions.

SPECIAL WASTE DISPOSAL FACILITIES

These are special purpose landfill sites for the disposal of chemical and hazardous wastes. The Federal Republic of Germany reported in 1977 on two such facilities: SCHWABACH and GALLENBACH.

The former is an extensive complex including special landfilling incinerator, detoxification plant, etc. The special landfill is completely sealed with an artificial sealer with a leachate collection system discharging to a waste treatment plant. Liquids and sludges are not accepted, but de-watered sludges such as metal hydroxide sludges are accepted.

The latter, Gallenbach, has a capacity of 1.4 million cubic meters. At the landfill incinerator residue, soils contaminates with oil, treated hydroxide sludges, processed wastes and special production wastes are accepted for disposal.

In 1977 the United States reported on a proposal to develop a demonstration landfill allowing for research work to be done on the following:

(a) Site selection methods
(b) Site preparation techniques
(c) Monitoring techniques
(d) Handling an operational procedure
(e) Costs
(f) social and institutional issues.

This site has still to be developed.

LANDFILL PRACTICE AND LANDFILL RESEARCH

In essence there are common threads running through all the landfilling experiences reported by the participating countries:

- A desire to protect the environment
- A desire to better understand the complexities of what has been, in the past, considered as a simple process.
- A desire to develop management tools for site selection, evaluation and management.

These can all be accomodated by pursuing appropriate research objectives and the development of information and technology transfer channels.

Research efforts in the past and in the future will be identified by the need to answer problems. It behooves the participating countries to find more answers by solving any one common problem only once. Interchanges of information through vehicles such as NATO/CCMS are essential.

This paper would be incomplete if it did not relate the landfill research work undertaken during the past 8 years regarding the landfill practices described earlier.

The frequently stated concerns and problems have been and remain:

- Environmental protection
- Leachate management
- Soil-waste interactions and
- Management of special wastes

Research activities in Phase I were very closely aligned with the pressing and common problems identified by participating countries in the landfill practice section. Research activities were directed to:

- The development of criteria and techniques for site selection and evaluation.
- The understanding of leachate its generation, behaviour and control.
- Developing an understanding of soil waste interactions.
- Development of alternatives to landfilling particularly special land disposal techniques including co-disposal.

In Phase I Canada contributed to research activities in all areas and continues to have an interest in all these elements.

Other participating countries were especially active in addressing those areas of common, but varying concern. For example:

The Federal Republic of Germany addressed - leachate problems
                                          - liner behaviour
                                          - gas generation
                                          - co-disposal
                                      and - special (but common) wastes.

France was investigating - problems with existing sites
                         - industrial waste disposal
                         - a controlled/monitored site

The Netherlands was investigating - leachate problems.

Norway addressed - leachate problems

United Kingdom undertook work on - site suitability
                                 - pollutant migration

U.S.A. - waste characterization and behaviour
       - pollutant migration
       - control technologies
       - alternative to landfilling

It should also be noted that findings of this research will be to some greater or lesser extent transferable. It is note worthy that within certain broad parameters the problems being addressed were universal in nature.

Phase II. Landfilling practice is not addressed in Phase II. However, some of the activities carried out under the auspices of Phase II and some of the findings could have a significant effect on landfilling practices.

Research activities that are of immediate interest are reported or include: Co-disposal - to establish the behaviour of selected industrial wastes when co-disposed with municipal refuse.
          Leachate - migration, attenuation, collection and treatment.
          Landfill gas - design and effectiveness of gas control systems.
          Waste Isolation - The use of liners, barriers and encapsulation.
          Immobilization - Solidification and stabilization

Alternative Methods - Land farming of hazardous waste.
Characterization and - To allow better prediction of gas and
　　　Description 　　leachate problem.
Technical criteria - Site suitability
　　for Landfilling 　Testing procedures
　　　　　　　　Landfilling Methods
　　　　　　　　Monitoring
　　　　　　　　Guidelines and codes of practices.

Additional research will be required as new waste streams are isolated or developed. More importantly it will allow prediction of inter-actions to be made before the fact. The fact of landfilling is established. The science of landfilling still has to be perfected.

SUMMARY OF CONCLUSIONS AND RECOMMENDATIONS

In essence a summary of the conclusions and recommendations of both Phase I and the Phase II reports as they apply to landfill practice are:

Landfill can be an effective way to dispose of certain wastes and is viewed as such by all participating countries. The chemical and biological mechanisms associated with landfilling-- the soil/waste and waste/waste reactions must be better understood. Also, research should be undertaken to examine the potential problems with micro pollutants (trace organics) and leachate collection and treatment.

Developing capabilities to predict the behaviour of contaminants, solids, liquids and gaseous must be a high priority to protect resources and public health.

The practice of landfilling must be improved, standardized within broad parameters and controlled by means of legislation, codes and guidelines.

Information exchange is essential between members of the scientific community, policy makers, countries and private citizens to promote the development and implementation of sound landfill control and prevention technology.

UNDERGROUND DISPOSAL AT HERFA-NEURODE

Dr. Gunnar Johnsson

Kali und Salz AG
Kassel, Federal Republic of Germany

INTRODUCTION

The Underground waste disposal facility Herfa-Neurode was established in 1972 by Kali und Salz AG, Kassel in compliance with the waste legislation of the Federal Republic of Germany.

The installation is integrated in the waste disposal scheme of the State of Hessen and is available for certain wastes from all states of the F.R.G. In the scope of the waste policy program, it accomplishes a substantial task with regard to the protection of the environment. In a few years, the underground waste disposal has become important also for other European countries. Since 1972, 270,000 tons have been disposed from several hundred producers. At present, 35,000 to 40,000 t/a are disposed from which about 25 % originate from foreign countries.

Underground waste disposal is assigned for such mainly water-soluble hazardous wastes, which cannot be disposed in other installations without creating a potential danger to water and/or air. This applies particularly if the biological or thermal decomposition is technically difficult or problematical with regard to environmental protection and/or too expensive. Although industry is improving its processes, works off residues and recycles wastes in a continuously growing amount, there are wastes left for which the underground disposal seems to be an optimal method.

During the last years, the tendency increased not to dilute and disperse the hazardous components of the residues or,

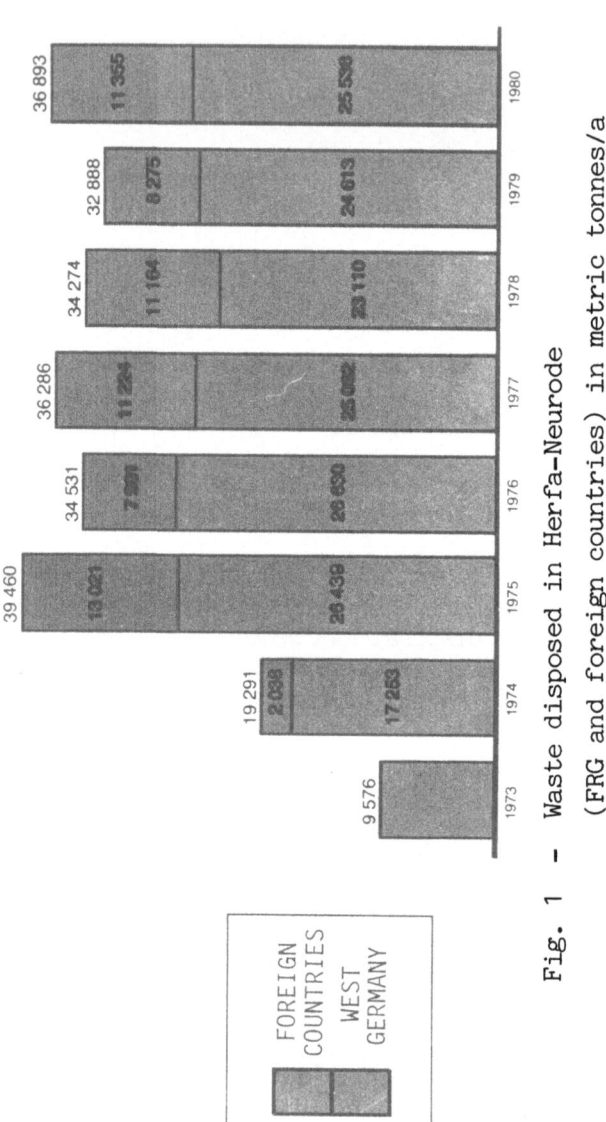

Fig. 1 — Waste disposed in Herfa-Neurode
(FRG and foreign countries) in metric tonnes/a

respectively, not to mix them with other materials and dispose
them in landfills. It seems to be much safer to seize those
hazardous components when and where they emerge, to seperate,
concentrate and treat them seperately and then remove them out
of the biosphere with reliable methods. In this regard, the
underground waste disposal plays an important part in the
completion of the disposal of hazardous wastes.

It is self-evident that such an operation has to be subject
to stringent operating standards, tight supervision by the
respective authorities and an exhaustive public information
policy to ensure public acceptance.

GEOLOGICAL, ROCKMECANICAL AND MINING CONDITIONS

The underground waste disposal is situated in the northern
part of the Werra-Fulda Salt Basin (Finkenwirth & Johnsson 1978,
Roth & Messer 1981).
Here the salt deposit, formed 250 million years ago, is between
250 and 300 m thick and lies in a depth of about 700 m.
It is covered by impermeable clay and shale layers, so that
it remained nearly untouched during the last 250 Million years.
In the Upper-Miocene, 15 Million years ago, the salt deposit
in some places was penetrated by basalt-intrusion. In spite of
the tremendous thermal and tectonic forces, the deposit remained
essentially unchanged. Together with the basalt-intrusions,
containing gases impregnated the salt in some places under high
pressure. The gases are still there and had no chance to escape
or otherwise change the location during the last 15 million years.
This has to be looked upon as an additional proof for the
imperviousness of the salt.

The rockmecanical properties and behavior of the salt-
deposit are thoroughly investigated and evaluated in the past
decades (Uhlenbecker 1971, 1974, 1978).Due to the mining-
method which has to be applied in the Werra-Basin the remaining
pillars are designed with such dimensions that the empty rooms
remain open and accesable for at least more than hundred years
(Messer 1978).

The waste disposal site is installed in the 1970-abandoned
section called Herfa-Neurode of the Wintershall-Mine. The active
mining sections are about 10 kilometers from the waste disposal
section. The ventilation system for the waste disposal section is
completely separated from the active mining area. Only the used
air goes to one of the two exhaust-shafts also used by the active
mining operation. This was one of the essential pre-requirements to
prove that the waste disposal operation will not affect the mining
operation. (NATO-CCMS - 1977) (Environmental Protection Technology
Series, 1975 and 1977).

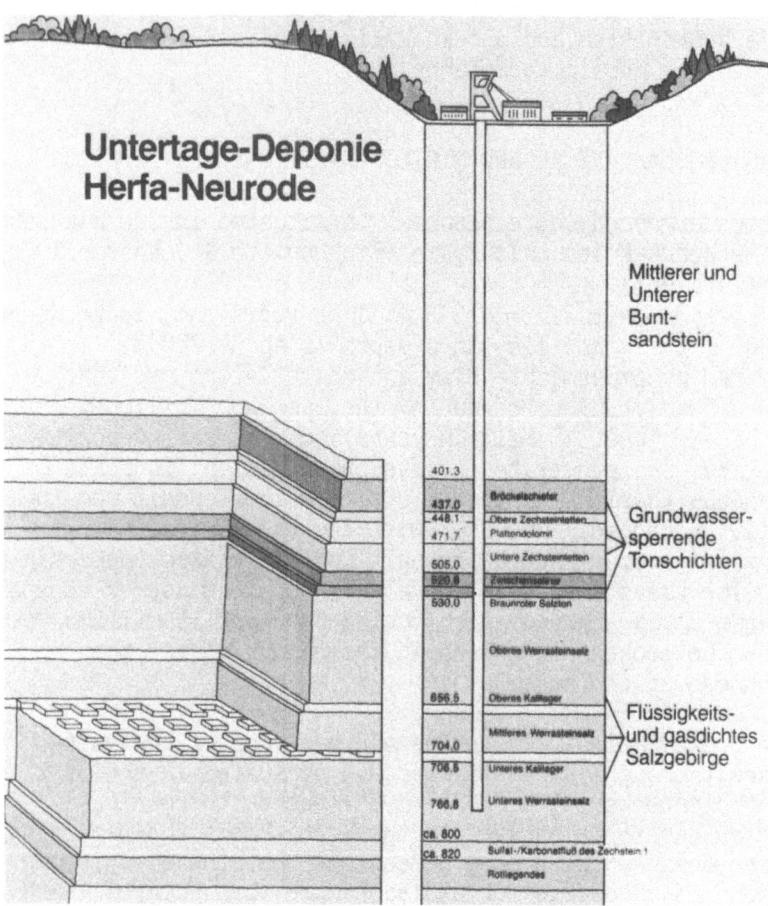

Fig. 2 - The geolocical situation

REQUIREMENTS, CONDITIONS AND REGULATIONS

In cooperation with the association of the German Chemical Industrie (VCI) and with the technical and legal experts of the competent authorities and chemical companies concerned, the general "business conditions" were developed in 1972. A special inquiry form "Formblatt A" is used for the definite identification of the wastes after inspections and discussions. The analytical determinations are performed together with the competent and responsible persons representing the waste producer.

To ensure the safety of the underground operation, the close cooperation with the waste producer and their familiarity with the processes used are imperative. Only the respective waste producer is able to be responsable for the necessary and sufficient declaration of the properties of the waste which are relevant for the operation of the underground facility. Different methods are applied in the course of the evaluation, depending on the characteristics and origin of the particular waste. Differential-thermoanalysis, gas-chromatography, determination of vapor pressure, ignitition-point, etc., complement the normal chemical analysis.

As an integral part of a mining operation, the underground waste disposal activities are supervised by the mining authority.

For each waste the "Formblatt A" is submitted for approval. The mining authority engages the Environmental Agency of the State of Hessen by filing a copy for review and agreement. If that agency has no objections, the approval is given by sending the signed "Formblatt A" back to the operation. Then the waste producer gets the "letter of acceptance" together with a copy of the approved "Formblatt A". For the purpose of transparent registration and documentation each waste has a code-number. With this number each drum or packaging is labeled.

The documentation in the "Inventory book" and in a special large-scale mining-map warrants the possibility that each waste disposed can easily be located at any time.

For wastes from foreign countries, an additional special procedure is required for the import licence, whereby the political implications in each case can be taken into account. The respective authority of the foreign country has to attest to the correctness of the declaration in "Formblatt A" and to confirm that in the respective country no equivalent feasibility to underground disposal is available. The import licence is valid for 3 years and then the whole procedure has to be repeated.

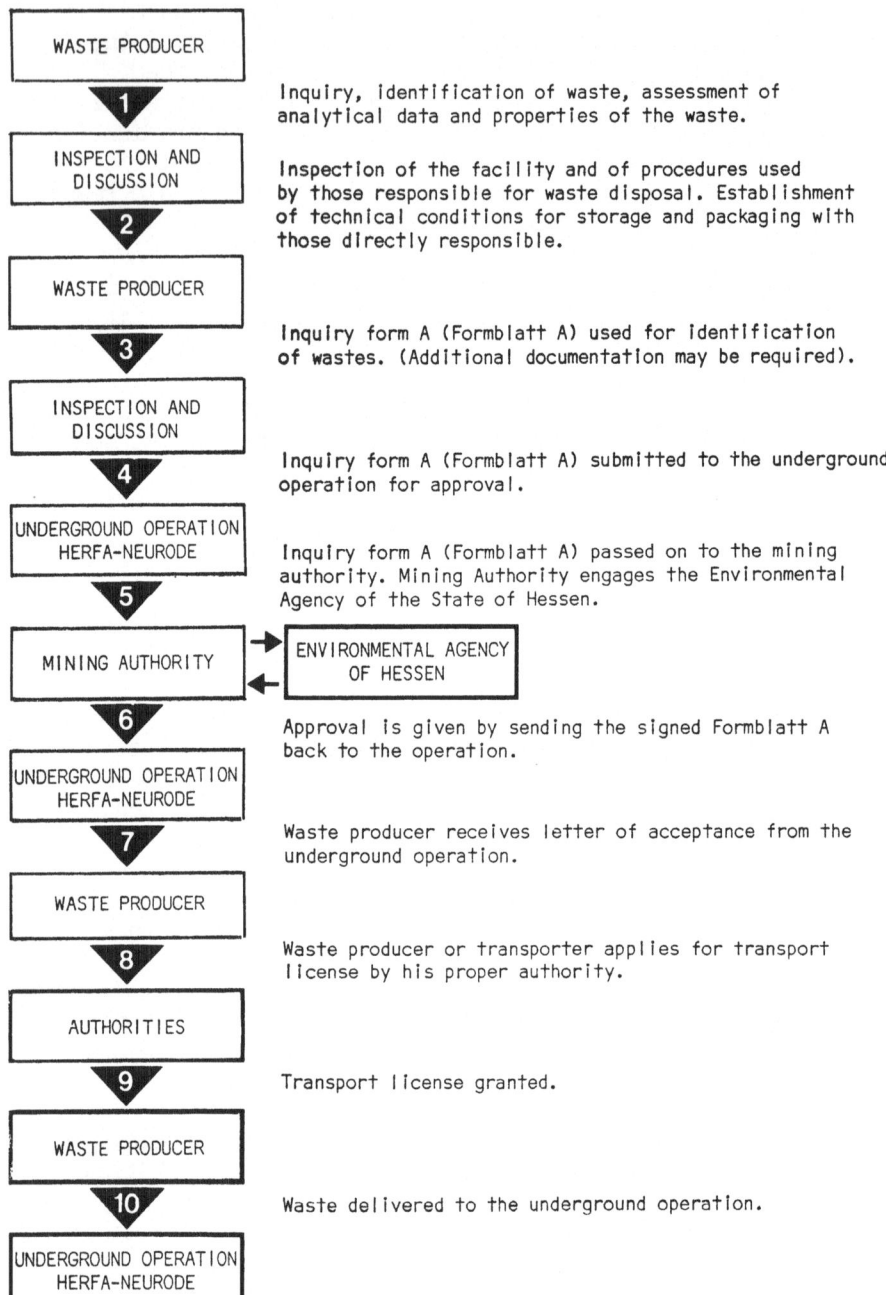

Inquiry, identification of waste, assessment of analytical data and properties of the waste.

Inspection of the facility and of procedures used by those responsible for waste disposal. Establishment of technical conditions for storage and packaging with those directly responsible.

Inquiry form A (Formblatt A) used for identification of wastes. (Additional documentation may be required).

Inquiry form A (Formblatt A) submitted to the underground operation for approval.

Inquiry form A (Formblatt A) passed on to the mining authority. Mining Authority engages the Environmental Agency of the State of Hessen.

Approval is given by sending the signed Formblatt A back to the operation.

Waste producer receives letter of acceptance from the underground operation.

Waste producer or transporter applies for transport license by his proper authority.

Transport license granted.

Waste delivered to the underground operation.

Fig. 3 – Procedure for acceptance and licence for each particular waste

GENERAL RESTRICTIONS

For safety reasons with regard to the underground operation, the following restictions are imposed:

(1)   No wastes are permitted which might develop under the prevailing conditions underground a gas-air-mixture which is explosive, ignitable, toxic or otherwise hazardous for the operation.
(2)   Wastes which might have a tendency to self-ignition or otherwise instability are prohibited.
(3)   No liquid wastes are permitted. If need be - and if other disposal methods are not available - those wastes must be transformed by appropriate methods mutually agreed upon in at least semi-solid consistency. In any case, it must be warranted that if damage to the packaging occurs, no free flowing liquids can escape and contaminate the surroundings.
(4)   No wastes are permitted which might react with the surrounding salt in such a way that the development of a danger to the operation cannot be excluded.

Wastes for which in the scope of these restrictions a disposal in the first instance is not possible, it might be necessary to apply methods of pre-treatment, conditioning, etc., agreed upon at the producer's plant to alter the relevant properties of the wastes. If this or changes in the processes from which the wastes originate are successfull, it might then be allowable to dispose of the waste. This procedure is already done from time to time in cases where the disposal with other methods is not available.

(5)   Wastes for which the possibility of a reaction with each other cannot positively be excluded have to be disposed at a sufficient distance so that they cannot come in contact with each other. This is possible due to the spacious extensions of the available section.
(6)   No wastes coming under the special regulations for radio-active materials are disposed at Herfa-Neurode.

SAFETY

For the underground operation,the regulations and rules for mining operations are in effect.

Additional injunctions and conditions are set with regard to the distinctive requirements of a hazardous waste disposal by the mining authority. Exceptional attention has to be directed to

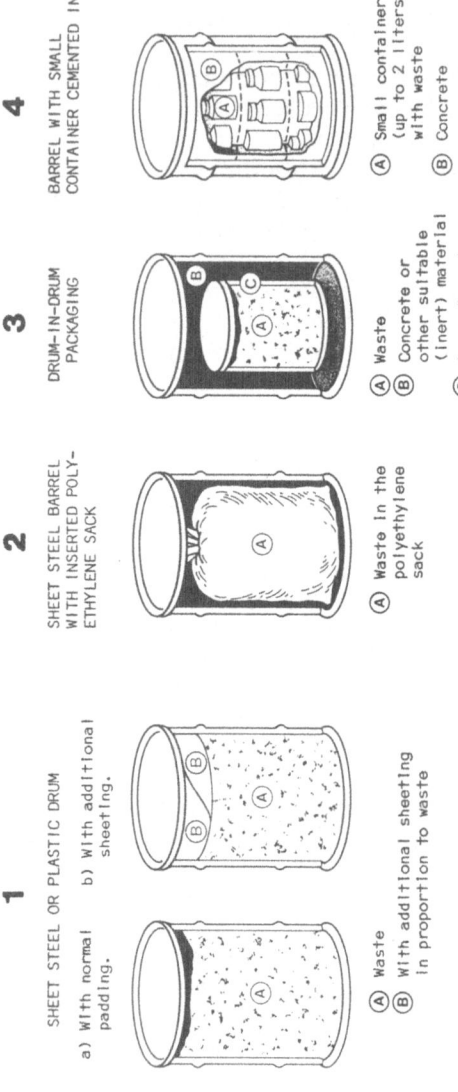

Fig. 4 – Examples for packaging of certain wastes

Fig. 5, 6 –

Examination and
Sampling of del-
ivered waste

Fig. 7 - Forklift unloading the truck

Fig. 8 – Inspection and control

monitoring for possible pollution of the underground atmosphere by
emissions due to the vapor-pressure of the wastes disposed. This is
done on a daily basis by testing with air test tubes, periodically
with the personal air sampler, and from time to time with special
equipment to detect and define even insignificant traces of polluting
components. Up to now there was no incident which affected safety
and working-conditions. This is due to the supply with ample amounts
of fresh air and the separation of the filled rooms from the ventil-
ation-circuit by brickwalls or saltdams. To prevent fire hazards,
all diesel-powered equipment is furnished with fire extinguishers
as well as the rooms where the actual disposal takes place. The
inert surrounding is another important safety factor.

CONTROL

Before acceptance of the wastes, the control document will be
executed. The documents accompanying the delivered wastes must conform to the declaration in "Formblatt A." Sampling is carried out
following a scheme defined by the authorities and management. The
samples are stored in a special room underground, where reference
samples are kept which were made available in the course of the
investigations with regard to the disposal permission.

If the control shows that there is something not clear,
management decides whether the particular delivery may be stored in
an underground control room until clarification or the delivery may
be refused. The waste producer is informed by telephone or telex
and must explain and clarify the reclamation before a final decision
is made. In a special book all objections are filed and, if advisable, pictures are taken for documentation.

Samples are taken regularly by the supervising authorities
too.

PACKAGING AND TRANSPORT

In general, the wastes are packed in 200 l -steeldrums which
have to be tightly closed. Depending on the properties of the
wastes, plastic drums may also be used. In some cases, special
precautions are applied, such as where offensive smells must be
prevented. Sometimes a "drum-in-drum" packaging is used. The space
between the drums is then filled up with concrete or other inert
material. This is done mainly in cases where wastes are in old
corroded drums and repackaging is not advisable. Care is taken so
that there is no outside contamination of the drums and palettes
by the wastes.

Fig. 9 – Erecting a brick-wall to shut of a filled room
from the ventilation circuit

Fig. 10 – Part of room where used surface treatment salts (hardening salts) from different producers are disposed

Fig. 11   -   Truck hauling retrieved waste to
                the shaft

For contaminated salts, plastic bags may be allowable.

For the purpose of safe handling, only one-way palettes are used. It is warranted that no person needs to come in contact with the drums.

The delivery is done mainly by trucks. But since March 1981, a railroad line to the mine Herfa-Neurode is in use, and it is expected that more and more railroad transport will be used in the future.

For transport and packaging the existing regulations for transport of hazardous materials are applied. New regulations for hazardous waste transport are being developed by the respective authorities. It is important to mention that the responsibility for packaging and transport remains with the waste producer. The devolution of responsibility takes place after unloading at the mine.

It must be emphasized that in general, the packaging is looked upon as means to allow safe and expedient mechanical handling and transport. For the evaluation of the long-term behavior of the wastes in the underground environment, the physico-chemical property of the particular waste is the primary distinctive feature and the basis for the decision whether a waste can be disposed or that permission must be declined.

THE WASTES DISPOSED

In the 9 years of operation more than one thousand different wastes are disposed. The greatest single group is about 10,000 t/a of surface treatment salts, which, because of their content of cyanide, nitrite, nitrate and varying contents of water soluble barium salts, are not to be disposed in open landfills without creating problems. This is of great importance for the users of those salts for hardening metal-surfaces. The fact that these residues can be disposed in an environmentally safe and economical manner enables industry to treat metal surfaces further with these salt compounds since alternative methods are not economically equivalent for all purposes. Experiments to work off these residues supported by the Federal government had no success. The costs for recycling were much higher than the commercial gain for the resulting products.

Another group of about 5,000 t/a are the high chlorinated solid residues from the chlorination of hydro-carbons. Also in this case, recycling is not economical due to the high energy consumption. A great number of distillation residues from color-production, pesticide-production and from the pharmaceutical

industry are disposed. Other wastes disposed are those containing
lead, chromium, cadmium, arsenic, selenium and other metals at
concentrations where recycling, at least in the time being, is not
economical. Also for a group of mercury-containing compounds, the
underground waste disposal seems to be an efficient solution, as
well as for PCB-containing wastes and impregnated insulation compounds.

If there are problems with disposal of hazardous wastes for an
industry, the underground waste disposal may offer a solution which
will be safe, efficient, and protect the environment. A detailed
consultation in any case is indispensable before the application is
filed for approval by the authorities.

RETRIEVING:

The special feature of this kind of waste disposal is that
the selective retrieving of certain disposed materials is
possible.

Up to now, more than 1,000 t disposed several years ago have been
retrieved on request by the waste producer and sent back for
further use.

In the planning of the operation, this is taken into account because
it may become more important in the time to come.

SUMMARY

(1) The extensive knowledge of the geological situation -
supported by 80 years of mining experience - permits the
statement that the wastes, deposited in the mined-out section
of this potash-mine are withdrawn from the biocycle for geological periods. Without human action they never will enter the
biosphere again.

(2) The underground waste disposal at Herfa-Neurode completes the
now existing available hazardous waste disposal facilities for
certain wastes which might cause environmental problems in
special landfills, chemical or thermal treatment plants.

(3) To warrant the safety of the underground operation, the
complete awareness of the relevant properties and the definite
identification of each particular waste is indispensable.

(4) In addition to the regulations already applied to hazardous waste management, a special procedure has been developed for examination, identification, approval, packaging, disposal and control.

(5) The producer of unavoidable hazardous wastes has to consider the special requirements of the underground waste disposal already in his production processes because he is the one who finally decides what kind of wastes emerge. Only he has the complete knowledge of the physico-chemical features of the wastes and is responsable for the declaration and method of disposal.
In many cases, this is now already done in the planning and licensing stage of certain processes.

(6) The capacity of the available space in the mined-out area of the mine allows long-term planning for the producer. If the operation of the underground waste disposal at Herfa-Neurode is managed successfully and in an environmentally sound manner, the operation can continue for years to come.

(7) A particular feature of this operation is the possibility of retrieving those wastes which might become a valuable matter in future times.

REFERENCES

ENVIRONMENTAL PROTECTION TECHNOLOGY SERIES (1975)
     Evaluation of hazardous wastes emplacement in mined openings
     EPA-600/2-75-040 - December 1975
ENVIRONMENTAL PROTECTION TECHNOLOGY SERIES (1977)
     Cost assessment for the emplacement of hazardous materials
     in a salt mine
     EPA-600/2-77-215 - November 1977
Finkenwirth, A. & Johnsson, G. (1978)
     Die Untertage-Deponie Herfa-Neurode bei Heringen/Werra
     Fifth International Symposium on Salt - Northern Ohio
     Geological Society (1978) 239 - 249
Messer, E. (1978)
     Die nordhessischen Kaligruben
     Kali und Steinsalz, (1978), Band 7, Heft 7, 806 - 318
NATO-CCMS-DISPOSAL OF HAZARDOUS WASTES (1977)
     The Underground Depot
     Report Nr. 69 (1977)

Roth, H. & Messer, E. (1981)
    Die Nutzung lagerstättenkundlicher Erkenntnisse für
    Planung und Betrieb der nordhessischen Kaliwerke.
    Kali und Steinsalz (1981), Band 7, Heft 7, 306 - 318
Uhlenbecker, F.-W. (1971)
    Gebirgsmechanische Untersuchungen auf dem Kaliwerk
    Hattorf (Werra-Revier).
    Kali und Steinsalz (1971), Band 5, Heft 10, 345 - 359
Uhlenbecker, F.-W. (1974)
    Neuere Forschungsergebnisse in der Gebirgsmechanik aus
    dem Salzbergbau.
    Kali und Steinsalz (1974), Band 6, Heft 9, 308 - 314
Uhlenbecker, F.-W. (1978)
    Neuere Forschungsergebnisse in der Gebirgsmechanik im
    Hinblick auf den Abbau von carnallitischen Kaliflözen.
    Fifth International Symposium on Salt - Northern Ohio
    Geological Society (1978), 413 - 422

DANISH HAZARDOUS WASTE SYSTEM

John Toffner-Clausen

Kommunekemi Ltd.
Nyborg, Denmark

PREVIOUS HISTORY REGARDING THE ESTABLISHMENT OF THE DANISH ABATEMENT SYSTEM

The Copenhagen area is the most industrialized area in Denmark, and it was therefore natural that it was also here that a solution had to be found to the problems caused by pollution originating from waste oil and chemical waste. A committee formed in the 1960s reached the conclusion that the first aim should be to set up a collection plant in Copenhagen and from there ship the waste to a Swedish company, AB Industridestillation in Stockholm, which was willing to receive and treat the waste. Later it was planned, in cooperation with a larger circle of municipalities, to consider the idea of establishing a treatment plant connected to the collection plant in Copenhagen.

Nevertheless, there was an awareness that the problem was not only a problem for the Copenhagen area but also for the rest of the country. As a consequence, the committee pointed out that the task before them was presumably a national task. The committee was, likewise, aware of the fact that special legislation in the field was a prerequisite to an effective solution.

However, there was a company, Danske Gasværkers Tjærekompagni A/S, founded in 1919, in Nyborg. Behind

the company stood a number of municipalities which operated coal gas works. The company produced tar as a by-
product of coal gas production and also produced agricultural chemicals for a number of years. Since the company was already owned by the municipalities and had a
number of facilities including a distillation plant and
tanks, etc., it was quite natural to expand those activities to include the treatment of waste oil and chemical
wastes collected from the whole country.

The municipalities became interested in the idea,
and support came from both industry and the state. Gradually, as the idea took shape, it was realized that a
task of this magnitude would have to be separated from
the Danske Gasværkers Tjærekompagni as an independent
company. The municipality of Copenhagen suspended its
previous plans on cooperation with a Swedish treatment
plant and supported the idea of a central Danish
treatment plant.

The prerequisites were now laid before the National Association of Danish Municipalities at their session in 1971 to vote for the establishment of Kommunekemi a/s, with all the Danish Municipalities, the
municipality of Copenhagen and of Frederiksberg plus
the Danske Gasværkers Tjærekompagni as stockholders.
Included in the plans for the solution to the problem
was establishment of a network of collection stations
throughout the whole country and a request to the Danish Parliament (Folketing) to pass the necessary legislation.

The request to the Danish Parliament was positively received and in May 1972, Parliament passed the Law
on the disposal, etc., of waste oil and chemical wastes.

The system as it is now known in Denmark has func-
tioned as an organized set of procedures since 1975 and
covers in a simplified form the following main components:

    - Legislation
    - Packaging systems
    - Transport systems

- Receiving systems (central collection stations and collection stations)
- Treatment plant, Kommunekemi

ACT ON THE DISPOSAL ETC. OF WASTE AND CHEMICAL WASTES

The Act and its application

The legal background for setting rules is found in Act No. 178 of May 24, 1972 on "Disposal of Waste Oil and Chemical Wastes," which is a framework act.

This means that the act does not contain definite provisions as to how waste oil and chemical wastes are to be disposed of etc., but gives the Minister administrative authority to issue regulations pertaining to waste oil and chemical wastes.

Such provisions are brought into effect by the Notifications on Waste Oil and Chemical Waste set forth by the Ministry of the Environment, cf. below.

The objectives of this act are to prevent and combat pollution caused by waste oil and chemical wastes. It covers every type of storage, transport or disposal, including burning or final depositing by industry as well as by the private sector.

The act and those rules contained in the act and those who have authority vested in the act may in principle issue regulations affecting both private citizens as well as public or private undertakings on whose premises waste oil or chemical waste arise.

The authorities

The Ministry of the Environment is the highest administrative authority within the field of environmental protection.

In practice, however, most of the central functions are carried out by the National Agency of Environmental Protection which acts as adviser to the Ministry of Environment.

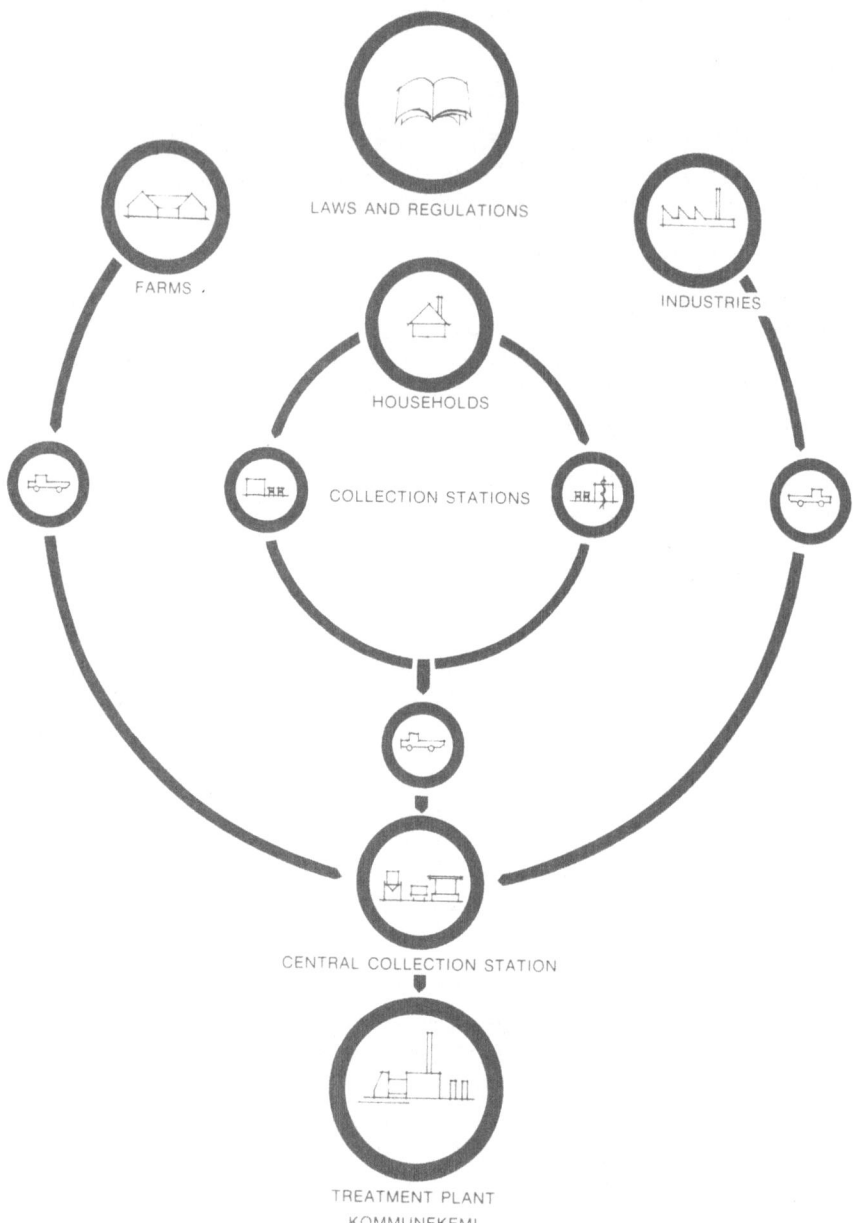

The Danish system for hazardous waste abatement

The Ministry of the Environment/the National Agency of Environmental Protection carry out duties pursuant to the act on the disposal of waste oil and chemical wastes consisting mainly of issuing general regulations plus making decisions in cases of complaint.

The administration of the Danish system for the disposal of waste oil and chemical waste is on a local plan organized such that the municipal council will take charge of the duties pursuant to the legislation on the protection of the environment.

The regional counsils (in Copenhagen, the metropolitan council) are obliged to supervise the municipal councils.

The municipal council in the individual municipalities are responsible for the establishment of local collection and receiving arrangements for waste oil and chemical waste.

Furthermore, it is the municipal councils which receive the notifications on waste oil and chemical waste, grant dispensation from obligatory delivery to the municipal collection arrangement, supervise that the regulations issued under the provisions of this act are observed, and, in the event of pollution, issue instructions as to the removal of the cause of pollution.

It is likewise the municipal council with its authority vested in the act which may take decisions on the collection of duties connected with the disposal of waste oil and chemical waste.

NOTIFICATION ON WASTE OIL AND CHEMICAL WASTE

Definition of waste oil

Waste oil is defined as oil products which are no longer deemed usable for their original purpose in their present state. In other words, oil products which have a market value or in practice constitute part of the usual turnover or use will also be covered by the regulations on the disposal of waste oil.

In principle, certain types of waste can be classified under both waste oil and chemical wastes. For

example, waste containing gasoline. As the concrete regulations in the two Notifications on the whole are similar this creates no difficulties in practice.

Definition of chemical waste

As it is not possible to define precisely what
chemical waste is, it has been decided to mention the
known types of chemical waste that arise in Denmark.The
selection has taken place after a preliminary analysis
of the structure and waste production of Danish industry. The selection was based on the assumption that an
inappropriate disposal etc., represented a health risk
and a risk to the environment because the type of waste
in question is corrosive, toxic or inflammable.

In an annex to the Notification on the basis of
this assumption, 50 types of wastes are listed which
are covered by the regulations of the Notification.

It must be expected that within industries producing
chemical waste, development of new processes or changes
in the present industry will occur and new types of
waste will appear while other types will disappear.
Chemical wastes not mentioned in the annex but having
characteristics similar to those mentioned in the annex
will therefore also be covered by the regulations in
the Notification on Chemical Waste.

Empty containers which have been contaminated by
chemical waste are covered by the Notification on
Chemical Waste. Municipalities can refuse to receive
explosive or spontaneously combustible chemical waste
and can order industry to render the waste non-explo-
sive or non-spontaneously comsustible before delivery.

Radioactive waste and waste from the production
and use of explosives and fireworks are not covered by
the Notification on Chemical Waste.

Note that in this respect earth which is polluted
by pure oil or chemicals will be covered by the Notification on Chemical Wastes.

As stated in the annex to the Notification, it is
the type of waste alone which determines whether the
waste in question is to be classified as chemical waste
and this covered by the regulations on the Notification
on Chemical Waste. No specification is given as to the

compound concentration for the waste. This is due to
the fact that the concentrations of the different compounds in a waste product may vary greatly because
waste cannot be a clearly defined product.

The local collection system,
(central and other collection stations)

In accordance with the regulations of the Notification, the municipal councils throughout the country
must set up a local collection system. A local collecting system includes a central collection station
covering a larger geographical area, usually several
municipalities, intended for the collection of waste
oil and chemical waste from industry. Also included in
the collection system is the establishment of collection stations covering a smaller geographical area,
usually a single municipality, intended for the collection of waste oil and chemical wastes from households.
It is important that the collection stations are situated centrally such that private citizens will have
easy access to them, for example by a local incinerator
plant, sewerage treatment plant or refuse dump, etc.

The personnel at both the central and other collection stations must have the necessary knowledge
about waste oil and chemical wastes. When receiving
this type of waste, the personnel at both the central
and other collection stations must check the packaging
and insure that is is safe and leakproof.

Such waste delivered to the collection stations are
sent via the central collection station to Kommunekemi
for the final disposal.

Statutory notice to industries

In accordance with the regulations of the Notification, industries in which waste oil and chemical
waste arise must give notice to the municipal council
concerning the waste.

An industry having less than 150 liters of waste
oil per year does not need to give notice to the municipal council.

The notice must contain information on the type,
quantity and packaging of the waste.

The statutory notification comes into effect, in
principle, when the chemical waste is produced. For industries producing waste which is continuously constant in quantity and composition, it is sufficient to
give notice of the waste for a period or "until
further notice." The industry only needs to renew the
notification in the event of changes occurring in the
quantity or composition of the chemical waste.

A by-product of the same type as one of the types
of chemical waste listed or which is like it, is not
covered by the statutory notification for delivery if
it is used in the present production process.

Statutory delivery to industries

Normally, the industry notification is followed
up by a statutory delivery of waste oil and chemical
waste to one of the locations selected by the municipal
council - a central collection station.

The municipal council can introduce a pick-up
arrangement in which the waste is picked up at the
various industries. If the municipal council has introduced such an arrangement, it will be easier for the municipal council to control whether the packaging and
transport occurs in an environmentally safe manner.

The rules for dispensation also apply at those
places where a pick-up arrangement has been established, cf. the following paragraph.

Dispensation

While it is not possible to obtain dispensation
from the statutory notification, the municipal council,
in accordance with the regulations of the Notification,
must grant dispensation from the statutory delivery if
an industry can prove that it is capable of disposing
of its own waste oil and chemical waste in an environmentally safe manner. When making a decision on dispensation cases, the municipal council takes into consideration whether the packaging, transport and the final
disposal of the waste is carried out safely. The municipal council must similarly grant dispensation from

participating in a pick-up arrangement if this is established.

Likewise, there is a possibility of obtaining dispensation from the statutory delivery for those industries desiring to transfer their waste oil and chemical waste to other industries for re-use or for final disposal.

If this be the case, the municipal council must control that the industry receiving the waste has obtained approval for the treatment of the chemical waste in question pursuant to the Environmental Protection Act, or in the case of waste oil, that burning or other treatment can be carried out in a environmentally safe manner. It must be mentioned that waste oil is often burned by industries not covered by the obligation to obtain approval.

According to the regulation, it is the industry itself which must prove that it has fulfilled the conditions necessary to obtain dispensation from the statutory delivery. If such proof is documented, the municipal council on the other hand must grant dispensation.

Chemical waste from the private sector

Private citizens are not obliged to give notice of their waste oil and chemical wastes, but are nevertheless responsible for the environmentally safe disposal of the waste.

As mentioned the Notification dictates that the municipal council must set up an arrangement insuring that all private citizens have a possibility for delivering waste oil and chemical wastes for disposal in an environmentally safe manner.

Special arrangements

Leftover medicines can be delivered to pharmacies, and dead mercury batteries to the distributor according to the special arrangements. In Copenhagen, waste oil and chemical waste from households can be delivered to paint shops.

PACKAGING, TRANSPORT AND COLLECTION OF
WASTE OIL AND CHEMICAL WASTES

General principles

Containers used for waste to be delivered at Kommunekemi must be solid, tightly sealed and, moreover,
be suitable for storage and transport. Further, for
some groups of wastes, the container is regarded as a
part of the waste and therefore the producer does not
receive compensation or reimbursement for the
containers. This is due to the fact that the containers
are often old and used and thus non-economical to clean
and reuse and partly because containers containing
solid waste or solid residues from liquid waste are
loaded into the incinerator.

Instructions regarding which type of container is
to be used for the various types of waste are given on
the chemical waste cards which must accompany the waste
during road transport.

Division of transports

The transport of waste oil and chemical wastes from
the waste producers to Kommunekemi can be naturally divided into two categories: Transport to the central
collection stations, (collection stations and central
collection station) and subsequently the transfer of
the wastes from the central collection station to Kommunekemi.

Transport from waste producer to the
central collection station

Transport of wastes from the collection station to
the central collection stations is administered by the
municipality. Often, a local private haulage contractor
handles the transports, or the transport is incorporated
into an established pick-up system.

Industries are obliged by law to deliver waste oil
and chemical waste, irrespective of the quantity. The
waste must be delivered to central collection stations
which are located no more than about 50 km away from
any point in Denmark.

Industries which produce larger quantities of
waste sometimes use their own truck or tank wagons. In

other instances, the wastes are removed in connection
with a return trip after a raw material delivery which
is handled by a haulage contractor. Usually the
transport of wastes is handled by special contractors
or by renovation companies. Large quantities of the
same type of wastes are often transported in gully
emptiers (oil and gasoline intercepter traps), in tank
wagons (liquid chemical waste), or in containers (pasty
waste, sludge, filter cakes).

For industries producing large quantities of
waste, transports can be carried out by rail, if the
industry in question has sufficient space for railway
tracks. If this be the case, tank wagons are made
available by Kommunekemi or by agreement with the
Danish State Railways.

Pick-up system

The collection methods mentioned above are based
on transport by the individual households or by the industry's own means of transport. In several respects
this is an unsuitable solution.

The National Agency of Environmental Protection
has also recommended the establishment of a public
pick-up system.

The advantages of such a system are:

- Easier control of transport equipment and
  packaging, insuring that the safety regulations
  are observed.

- A pick-up system seems at times to be more
  effective than a delivery system efficiency-
  wise, when waste quantities per year per
  employee is used as a gauge.

Protection of the environment is thus best managed
through a pick-up system.

Often it is economically advantageous for the industries to participate in a pick-up system as the
routes can be planned rationally, making transport
costs more attractive. This can be further supported by
the fee and tariff policy used by the municipalities
involved.

The pick-up system quite often encompasses the same municipalities which have collectively established and run a central collection station. The administration and arrangement of transports are handled by the municipality in which the central collection station is located. In practice, contact with the waste producer and with the haulage contractor is managed by the personnel at the central collection station which, administration-wise, comes under the technical administration department of the municipality.

Transport from the central collection
station to Kommunekemi

Transport of waste from the central collection station for waste oil and chemical waste to Kommunekemi is carried out by rail. However, for a few individual stations (those not supplied with railway tracks), the waste must first be transported to a reloading site in the immediate vicinity of the railway net.

Central Collection Stations
Location and planning

At the delegation meeting in 1971, held by the National Association of Danish Municipalities at which it was decided to found Kommunekemi, it was also decided to set up local collection arrangements for waste oil and chemical wastes. This decision was later followed up by legislation called the "Notification on Waste Oil," as well as the "Notification on Chemical Wastes" rules that local collection arrangements must be established. A closer description is given of the collection arrangement in the circulars issued by the National Agency of Environmental Protection pursuant to the two Notifications. A receiving arrangement is most often based upon the central collection station which receives wastes from a local area. This can be a district or part of a district. The arrangement can be supplemented with a pick-up system plus a number of the above-described collection stations, at least one per municipality. The central collection station receives waste from industries, agriculture and from collection stations, while collection stations receive wastes from private households.

At present, 21 central collection stations are located all over the country in the following cities:

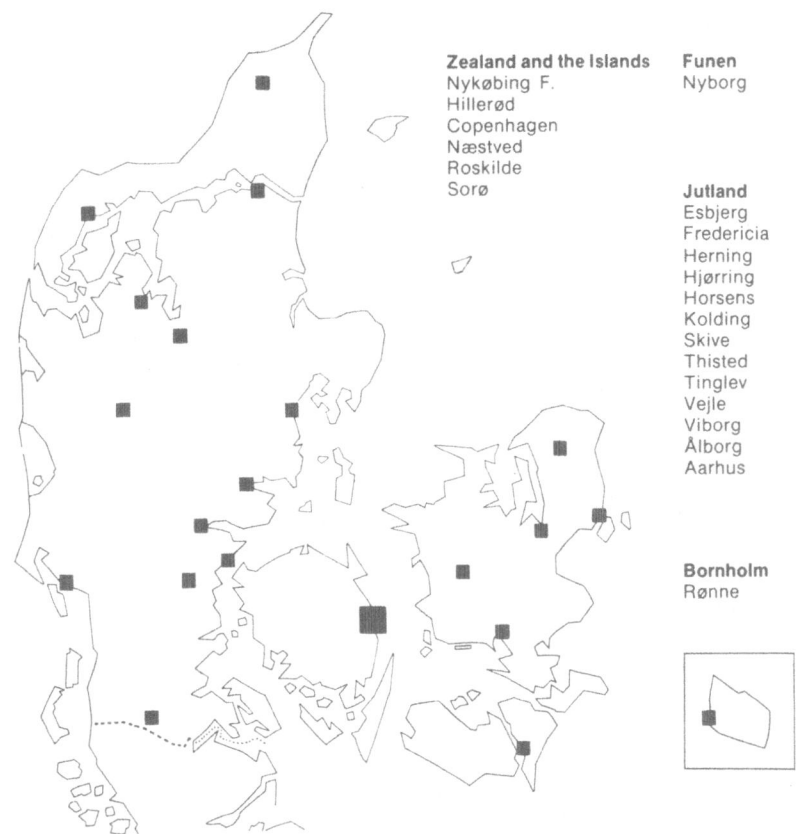

Zealand and the Islands
Nykøbing F.
Hillerød
Copenhagen
Næstved
Roskilde
Sorø

Funen
Nyborg

Jutland
Esbjerg
Fredericia
Herning
Hjørring
Horsens
Kolding
Skive
Thisted
Tinglev
Vejle
Viborg
Ålborg
Aarhus

Bornholm
Rønne

Location of central collection stations

　　In most cases a central collection station is
established through cooperation between the municipalities of an area and is run by a board which is elected
by these municipalities. Expenses connected with the
establishment and operation of a central collection
station are usually distributed between the municipalities according to their number of inhabitants or according to a distribution scheme by which the population is rated more or less according to whether it is
an agricultural municipality or a more industrialized
area.

When the number and location of the central collection
stations had to be laid down, several considerations had
to be taken into account. Each central collection
station had to have a sufficiently large industrial
base to make possible full benefit from its facilities
and personnel, and the transport distance to the

central collection station had to be evaluated with an
eye on cutting down transport costs for the industries.
It is possible to obtain an optimum economy when a
central collection station is supplemented by a pick-up
system as is the case for several central collection
stations, even if the transport distance is increased
due to a more efficient use of the capacity at the
central collection station.

Interiour set-up of a central collection station

Transport from the central collection station to
Kommunekemi is mainly carried out by rail. For several
reasons, this form of transport was chosen:

1. Rail transport must be considered as the safest form
   of transport available.
2. Rail transport is practical and economical, as
   freight cars stationed at Kommunekemi, as well as at
   the collection stations, are only connected with
   reasonable costs.
3. Kommunekemi's railway tank wagons have a special
   advantage, which is mentioned below.

To a great extent, two conditions determined the
location of the central collection station: Whether the
location in question had railway tracks which could be
used by the central collection station, or whether a
railway track could be laid at a reasonable cost.

A central collection station is a fenced-in area
having a railway track. On the track there is room for
2-4 railway cars. The railway cars are owned by Kommunekemi (in all 115). Along part of the railway track, a
ramp is built having at one of its end a covered
storage ramp for packaged inflammable wastes and at the
other end a covered storage ramp for packaged non-in-
flammable toxic wastes. When a sufficient amount of
packed waste has been received to fill a freight car,
such a freight car is requested and loaded at the
ramp.

Tank wagons and gully emptiers arriving at the
central collection station containing liquid wastes are
emptied at a specially designed emptying site and the
waste is pumped or moved by a stationary pipe into the
stationary railway cars at the site.

Wastes from oil and gasoline intercepter traps
which, as a rule, contain a major part of water and are

easily separated into an oil phase and water phase, are pumped into a specially designed tank in which separation takes place by settling. From the settling tank, the water phase can later be discharged into the central collection station's sewer system which is supplied with an oil intercepter trap.

Often when a gully emptier is emptied of its liquid waste, a residue of sludge is left in the tank.

Therefore, at the central collection station a concrete pit has been constructed where the gully emptier's non-pumpable sludge residue can be expelled together with the residue of the liquid load. From the pit, the liquid waste is pumped into railway cars with the aid of a submerged level regulated whirl pump, while the non-pumpable waste is lifted from the bottom of the pit by a spiral conveyor emptying into a clamp lid cask.

Other installations at the central collection station are an office building, a weighbridge, a shed for various tools and a place for return containers.

The declaration system

To insure that Kommunekemi can treat wastes in an environmentally safe manner and that emissions levels set by the authorities are observed, sufficient information must be made available on the type and characteristics of the waste. Therefore every waste supplier must complete a declaration as long as their wastes are received at the central collection station. station.

The declaration is supplied with a serial number in the right hand corner. Directly under this number is a space to write the number of the chemical waste card in question. In Space 2, seven main groups are listed in which the waste is to be placed. The supplier must indicate the main group to which the waste belongs. The respective letter and the serial number of the declaration must be written on all the containers containing the respective waste (a declaration must be completed for each type of waste) so that the personnel at the reception department at Kommunekemi can see to

which main group the waste belongs and, thus, which
treatment it is to undergo.

A = Mineral oil waste
B = Halogenous solvent waste and pumpable, halogenous
    or sulphurous organic chemical wastes
C = Solvent wastes
H = Organic chemical wastes, halogen- & sulphur-free
T = Pesticide-containing wastes
X = Inorganic chemical wastes
Z = Non-pumpable halogenous or sulphurous organic
    chemical wastes and other wastes

In Space 3, the charasteristics and components of
the waste are given plus the process from which the
waste is derived. The last item of information is
important to the personnel at Kommunekemi when
evaluating the information already available on the
state and components of the waste.

In space 4, the quantity and mode of delivery are
given, including the number and type of container. In
Space 5, the name, address and telephone number of the
industry is given plus the name of a person within the
industry who, if necessary, can give further
information about the charasteristics and composition
of the waste.
Finally, the declaration is signed by the supplier
who is held responsible for the correctness of the
given information and is also liable for payment of the
charge connected with the treatment of the waste.

The last space on the declaration is for use by
the central collection station only. In this space the
number of the railway tank wagon or freight car is
written and which containers have been handed over to
the supplier or possibly exchanged. Finally, the
personnel at the central collection station signs the
receipt for the receipt of the wastes and hands the receipt over to the supplier.

The declaration has 5 copies, and on the back of
copies Nos. 2 and 3, spaces are reserved for use by the
laboratory personnel at Kommunekemi and by the chemical
engineer who processes the declaration and gives information on the examinations performed at the laboratory
and the conclusions drawn regarding the waste.

Furthermore, the back of copies Nos. 2 and 3 are also used for calculating the treatment charge, in part on the basis of the information given from the laboratory.

Attached to the declaration are instructions on how to complete the declaration. Under the instructions are also given the conditions which must be fulfilled before the waste can be received.

THE TREATMENT PLANT, ITS FACILITIES AND APPLICATION

The receiving plant

The environmentally dangerous waste is transported to Kommunekemi by either road or rail.

In 1980, Kommunekemi received about 75% of all environmentally dangerous wastes by rail.

Connected to rail transport, railway tank wagons and railway freight cars are used. Railway tank wagons are used for transporting, liquid wastes while freight cars are used for transporting both liquid and solid packaged wastes.

A KK railway car

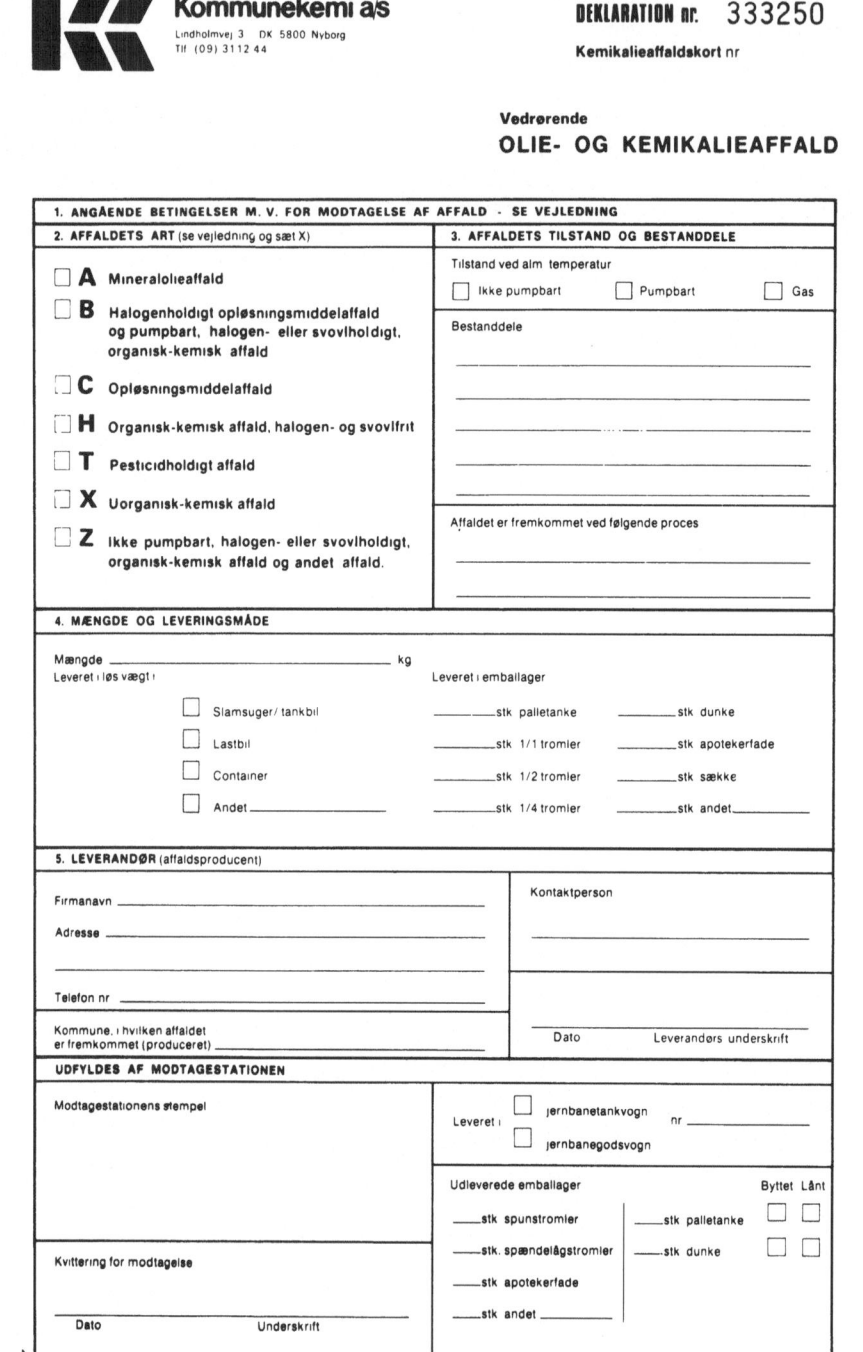

Front page of declaration

Kommunekemi receives both liquid and solid wastes by road. For the transports, tank wagons, gully emptiers, trucks and container trucks are used.

When the wastes arrive at Kommunekemi, they are distributed to the following receiving sections, some of which will be described in greater detail in the following:

Railway tank wagon emptying section
Drum emptying section
Inorganic chemical waste section
Receiving pit for waste oil
Receiving pit for burnable wastes
Container site

When Kommunekemi was to plan and project its receiving section, i.a. liquid wastes, careful considerations were made as to which system was to be used for the emptying operations. A vacuum system or pumps could be used for the emptying of tank wagons and containers holding liquid wastes.

As in many cases liquid wastes contain solid impurities, combined with viscosities varying from easy-flowing to the consistancy of paint, it was decided to combine the two systems. The system is combined in such a way that the waste is sucked into a vacuum tank. When the tanks are full, the vacuum seal is broken by atmospheric air after which the contents are transferred to the respective plants by using a so-called whirl pump. In the case where the waste does not flow towards the pumps, the tanks can be pressurized.

Railway tank wagon emptying:

In the railway tank wagon emptying section, two 30 m^3 vacuum tanks, which are evacuated by rotation vacuum pumps, are located. Waste is transferred from the tank wagons and gully emptiers to one of the two vacuum tanks.

When all the waste has been sucked out, the transport tanks are cleaned with jet water and this water is also transferred to the vacuum tanks and mixed with the waste.

When the vacuum seal has been removed from the
tanks by admitting atmospheric air, all the wastes are
pumped to the respective treatment plants and a sample
is taken during pumping by automatic sampling
equipment.

A heating system is built in the railway tank
wagon emptying section which can be connected to the
heating coil of the railway tank wagons. The system is
used in connection with wastes which become fluid by
heating, for example, heavy fuel oil.

Neutralized or alkaline inorganic chemical wastes
delivered in tank wagons or gully emptiers are emptied
by using pumps.

Drum emptying section:

Five 10 m^3 vacuum tanks, which are emptied by
using rotation vacuum pumps are placed in the drum emptying section. The liquid waste is sucked from the bung
drums and other containers into one of the five vacuum
tanks.

Those drums which can be emptied are transferred
to a drum storage where a scrap dealer takes over. The
drums are pressed into blocks and sent to the steel
rolling mill.

Drums containing residue of wastes are transported
by conveyor belt to "cutting up" where one of the end
lids are removed.

In the drum cutting section, the open drums are
covered by plastic to reduce the emission of volatile
substances. No container must contain more than 100
liters of waste when leaving the drum cutting section,
in regard to later incineration.

The drums are transported from the drum cutting
section to either the incinerator plant or to one of
the two storage halls.

The liquid wastes which have been sucked out of
the drums are pumped to the respective treatment
plants.

Storage plant

At Kommunekemi, there are storage plants for liquid as well as solid wastes dangerous to the environment.

For mixed waste oil, that is, a mixture of oil and water, there is a 3,500 m^3 tank.

For chemical wastes containing solvents, a tank capacity of about 4,000 m^3 is available of which 400 m^3 is underground tanks.

The underground tanks are used especially for halogenous wastes which gives added protection in the event of fire. Halogenous wastes can emit toxic gases when heated.

All tanks containing environmentally dangerous wastes in inflammable Gard I are supplied with inert gas. Nitrogen is used as inert gas at Kommunekemi and the supply comes from a central tank containing liquid nitrogen.

The waste is pumped from the storage tank in elevated pipes to the incinerator plants.

For solid packaged waste Kommunekemi has two storage halls of about 2,000 m^3. The storage halls have a closed drainage system.

Waste oil plant

In the waste oil plant, the waste oil is cleansed of impurities, producing a fuel oil which can be used in industry.

From the railway tank wagon emptying section and the drum emptying section, oil wastes are pumped to the receiving tanks and heated to about 90° C. A number of volatile substances evaporate at this temperature where gasoline is the main element. The gasoline gases are concentrated in a cooler and the condensate is collected and incinerated at a later time.

The oil sludge is collected at the bottom of the receiving tanks during heating. This sludge is transported by gully emptiers to a tank by the incinerator plat from where it can be transferred to the rotary kiln in the plant by using a special dosage pump.

The waste oil which now consists of a mixture of contaminated oil and water is pumped through a filter station to a large storage tank. In the storage tank, a separation occurs in an oil phase, a water phase and a precipation phase of light sludge.

The oil phase is transferred to a treatment tank where the oil is heated to a temperature of about 110°C. By this a further distillation of volatile substances and water takes place. Likewise, a watery sludge phase is separated at the bottom of the tank.

The oil phase finally passes another filter station before it is ready to be used as a fuel oil.

Gradually, the water from the storage tank is transferred to a tank for contaminated water, from which it is pumped through an elevated pipe to the incinerator plant where it can be injected into the after-burning chamber.

The present plant has a capacity of 15,000 tons of waste oil per year, from which about 5,000 tons of fuel oil are produced.

At present the waste oil plant is under expansion which will increase the annual capacity to about 25,000 tons. Moreover this new plant will be able to receive solid waste originating from spill accidents as this type of waste is likewise processed to fuel oil.

The inorganic waste which is not often received in railway tank wagons or gully emptiers, arrives in plastic pallet tanks holding about 800 liters or in plastic cans holding about 30 liters.

These containers are emptied at the receiving station at the inorganic plant by using pumps, and the wastes are transferred to the storage tanks.

The various types of chemical waste are mixed in 3 main groups:

Alkaline cyanide-containing wastes
Acidic chromate-containing wastes
Acidic iron-containing wastes

Moreover, hydrofluoric acid-containing waste is collected separately in a polyethylene tank as hydrofluoric acid attacks the ceramic lining of the storage tanks.

Treatment plant for inorganic chemical wastes

The treatment is based on the conventional detoxification process, where the process is carried out in
stages starting with an oxidation of cyanide using a
hypochlorite solution next, a reduction of chromate
using iron-containing (ferro) wastes, and finishing with
a neutralization process using milk of lime.

The heavy metal content of the waste is
precipitated in the form of hydroxides which can be
filtered off in a filter press chamber. The filtrate
can be discharged into the municipal sewer system after
a neutralization process.

The filter cakes originating from the press are
collected in a container and transported to a controlled landfill site for deposit.

Hydrofluoric acid-containing wastes are detoxified
through a special (stage) process. After the process is
completed, the waste can be mixed with the rest of the
detoxified and neutralized wastes.

The plant is prepared for a capacity of 30,000
tons per year and can treat about 15,000 tons a year
with one filter press chamber.

Treatment plant for organic chemical wastes
Pumpable halogenous solvent wastes:

This type of waste is burned in Incinerator II.
There is no recovery of the heat produced during the
burning process. The incineration is carried out at a
temperature of about 1000° C, at which temperature the
halogenous content (mainly chlorine) of the waste is
converted to hydrogen halide. The hydrogen halide is
washed out of the flue gases in a scrubber using water.
The weak hydrochloric acid solution is neutralized by
using a sodium hydroxide solution. From the scrubber
the flue gases pass through a venturi scrubber which
removes most of the dust content of the flue gas.

The wash-water from the venturi scrubber is acidic
and must be neutralized. The wash-water's dust content
and the sludge content formed during neutralization are
removed by a two stage sedimentation process. The main
part of the wash-water is recirculated to the scrubber
and venturi scrubber, while the precipitated sludge is
dewatered in a filter press chamber.

The burning capacity of the plant is approximately
1.5 tons of solvent wastes per hour using about 100
liters of support fuel.

Liquid and solid organic chemical waste containing a low sulphur and halogen content

These wastes are burned in a rotary kiln plant (I)
with a boiler and electrostatic filter. The wastes are
loaded into the rotary kiln which has a temperature of
1200-1300° C, at which temperature a fluid slag is
formed.

The liquid waste is conveyed to the rotary kiln
via lance-type nozzels while the solid waste is either
loaded as bulk weight or in packages.

The ratio between liquid and solid waste is as
3:1.

The fluid slag is cooled in a water bath and is
collected in a container. The slag is deposited at a
controlled landfill site.

In the after-burning chamber, a total burning off
of the flue gases takes place at a temperature of about
900° C.

Contaminated water, mostly originating from the
waste oil treatment plant, is injected into the after-
burning chamber achieving in part a burning off of the
impurites found in the water and in part a lowering of
the flue gas temperature before entering the boiler
section. The flue gas is cooled down from about 700° C
to about 260° C in the boiler. The boiler produces
approx. 17 tons of steam per hour at a pressure of
about 11 bar. About 20% of the steam quantity is used
at Kommunekemi, while the remainder is sold to Nyborg
for the district heating system.

The flue gas is cleaned for dust through an electrostatic filter before the flue gas via a suction
funnel blower is let out through a 60 m high stack.
The dust from the plant is likewise deposited at a
controlled landfill site.

The incinerator plant has a capacity of 5 tons of
organic chemical wastes per hour and 2 tons of contaminated water.

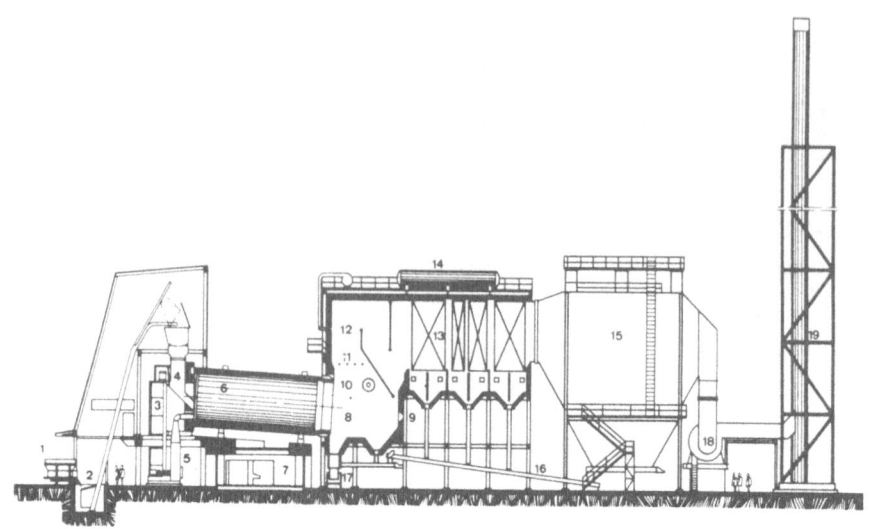

The incinerator plant I

1. Lamellar conveyer
2. Bucket elevator
3. Borrel feeding device
4. Feeding chute
5. Combustion air
6. Revolving drum
7. Control room
8. After-burning chamber
9. Solvent burner
10. Used oil burner
11. Secondary air
12. Tertiary air
13. Cooling boiler
14. Boil drum
15. Precipitator
16. Ash separation
17. Clinker channel
18. Fan
19. Stack

Water plant

As previously mentioned, all contaminated water
from the waste oil plant is burned in the big incinerator plant. Wastewater from the treatment plant for inorganic chemical waste is sufficiently cleaned in the
treatment plant itself, so that it can be discharged
into the municipal sewer system after having passed
through a clearing basin.

At Kommunekemi, all surface water is passed through an oil separation basin where possible oil etc. is retained and sludge is precipitated.

Before the waste water leaves Kommunekemi, it is passed through a monitoring station where the flow, pH value and temperature are monitored and registered. At the same time a sample of the wastewater proportional to the flow is taken by an automatic sampling equipment

Depositing site, Klintholm

Kommunekemi has a controlled landfill site about 20 km south of Nyborg. The site ist located in a area where agricutural lime is excavated.The site is divided into 3 sections, one for filter cakes originating from the inorganic plant, one for slag and ashes from the incinerators, and one for other wastes.

The section used for filter cakes and sludge/ashes is covered with a plastic membrane to reduce the amount of percolate.

The landfill site is located about 200 meters from the coast of the Great Belt and all percolate from the site flows with the ground water to the sea. The site covers an area of about 5 hectares.

The authorities demand a tight control of the end-products deposited from Kommunekemi and 3 time a year samples are taken for analysis of the percolate, sur-face water and drilling water.

Kommunekemi receives about 500 tons annually of environmentally dangerous wastes which can not be treated at the plant in Nyborg. The wastes can, for example, be hardening salts or wastes containing mercu-ry compounds.

Kommunekemi has made an agreement with a German company, Kali und Salz AG, which has been granted permission by the German authorities to deposit this type of environmentally dangerous waste in abandoned salt mines lying about 700 m under the surface of the earth.

Laboratory and spare parts depot

Kommunekemi's laboratory takes care of three main task:

1. To control the incoming wastes
2. To carry out analysis of the operations connected
   with the treatment of the wastes.
3. To control Kommunekemi's drainage facilities and
   carry out controls of the landfill site.

   The task of controlling the incoming wastes is
very important as this, together with the information
given by the waste producer on the waste, form the basis
for an environmentally correct treatment of the waste.

Future projects

   In the period between 1980-1982, Kommunekemi will
double its incineration capacity for organic chemical
wastes as yet another rotary kiln plant is under construction.

   Above and beyond this, Kommunekemi does not have
for the time being any plans on starting major plant
construction for the adaption of treatment methods.

COST ESTIMATES

Investments:

   Kommunekemi was constructed in stages.

   - Construction Stage 1 covers the receiving plant
     and administration building, etc. This stage
     amounted to about Dkr 65 million and was con-
     structed between 1973 and 1975.

   - Construction Stage 2 covers the enlargment of
     the receiving plant and storage plant plus the
     building of a smaler processing plant. The in-
     vestments connected with Construction Stage 2
     amounted to about Dkr 20 million and was con-
     structed between 1976 and 1977.

   - At present, Construction Stage 3 is in progress
     at Kommunekemi. This stage covers among other
     things a new rotary kiln plant, a new waste oil
     plant, and a new staff building, etc. Construc-
     tion Stage 3 is planned to be completed in 1982,
     with a total investment amounting to about Dkr
     80 million.

The new incinerator plant III under construction

At the end of 1982, the total investment capital invested in Kommunekemi's treatment plants in Nyborg will amount to approximately Dkr 165 million.

Income, expenditure, results

A simplified budget for Kommunekemi based on the 1980 accounting results, looks like this:

Incoming expences etc.	39,913
Sale of merchandise	1,200
Sale of packages and scrap	180
Gross turnover	41,293
Production & operating costs	24,974
Assets	16,319
Administration costs	4,553
Profits	11,766
Financing incomes	6,310
Primary results before depreciation	18,076
Ordinaire and extra ordinaire depreciation	12,993
Results	5,083
	=====

Capital expenditures

Stock capital in Kommunekemi is Dkr 1 million.

That means that all investments have been carried out on loans.

All the Danish municipalities together put up the loans and the first loan was effected in 1973.

The loans are given on very favourable conditions as they are interest and repayment free until January 1, 1983.

In 1983 negotiations will be taken up with the lenders (the municipal VAT fund) concerning the conditions for repayment and interest rates for the loan capital.

In other words, Kommunekemi has not had capital expenditures for the first 10 years of its exsistence.

Depreciation policy

It has been decided to depreciate the buildings over a 15 year period and the machines over a 10 year period.

DU PONT CHAMBERS WORKS - INTEGRATED SITE WASTE DISPOSAL SYSTEM

T.E. Lewis

E.I. du Pont de Nemours & Company, Inc.
Deepwater, New Jersey, United States

SUMMARY

At the Du Pont Chambers Works, continuing efforts are being made to reduce the generation of waste materials. At the same time, recognizing the inevitable continuation of need for adequate treatment and disposal, we are working to make our waste management system as efficient and reliable as possible by applying new technology as it becomes available.

INTRODUCTION

This paper discusses an integrated waste disposal system for a large, diverse chemical manufacturing site. The facility is the Du Pont Company's Chambers Works, which is located in southern New Jersey on the Delaware River estuary. Chambers Works occupies approximately 500 acres and employs almost 5,000 employees. Manufactured products include; "Freon" fluorocarbons, a variety of substituted aromatic compounds, petroleum additives, paper and textile-treating chemicals, elastomers, mineral acids, and miscellaneous other chemicals.

Even though the plant utilizes many opportunities to recycle byproducts, the manufacturing operations generate significant quantities of waste (Fig. 1) in the form of liquids and vapors, dilute aqueous waste water and solids. For maximum environmental control, the disposal facilities for these wastes have been integrated so that they supplement each other.

345

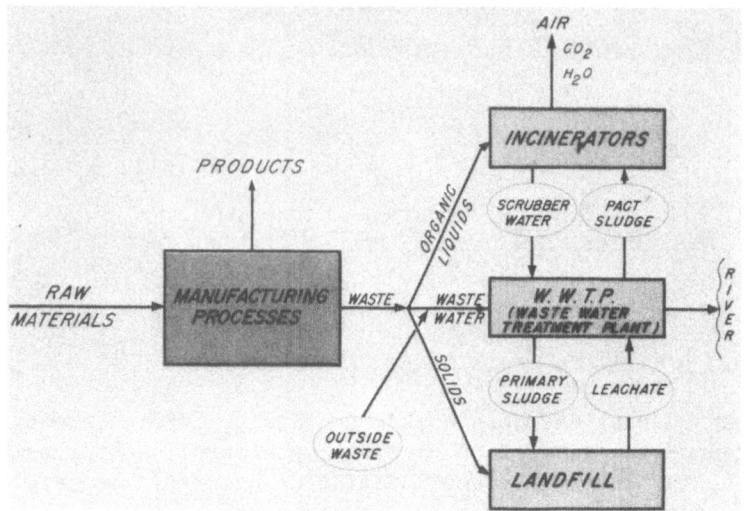

Fig. 1.   Chambers Works Waste Management - Schematic

DISCUSSION

Site Facilities

On site are three disposal facilities - incinerators, secure landfill, and waste water treatment.

Incinerators

Combustible organic liquid wastes are fed to the liquid chemical incinerator (Fig. 2) at an average rate of over 1,000 pounds per hour. It is necessary, of course, to provide the proper blend of compatible wastes to satisfy: (1) the physical and combustion characteristics desired for smooth, efficient operation of the incinerator and: (2) the requirements for necessary cleaning of combustion gas before emitting to the atmosphere. The main combustion chamber is fire-brick lined and provides a volume of over one thousand cubic feet for residence time of about two seconds at 2,500°F to assure destruction of organic compounds. Since many of these wastes contain halogenated compounds, two stages of water scrubbing are provided with the acidic scrubber water being sent to the Waste Water Treatment Plant (WWTP) for neutralization.

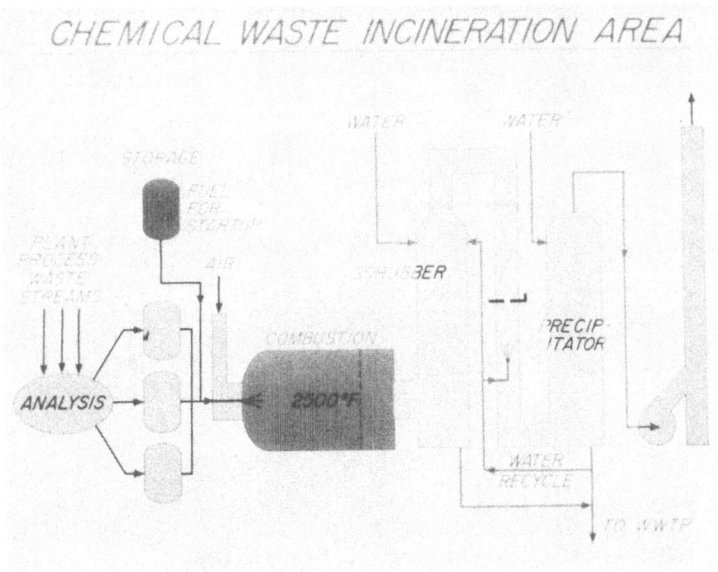

Fig. 2.   Liquid Chemical Waste Incineration Facility
          Chambers Works, E. I. du Pont de Nemours

     Some waste streams introduce very fine solid, inorganic materials
to the combustion gas stream, which pass through the water scrubbing
stages and are removed by a final irrigated, electrostatic precipitator whose water stream and contained solids also are sent to the
WWTP.

Secure Landfill

     A second category of waste is the chemical solids which are
not ignitable, nor are they soluble in water to permit treatment in
the WWTP. These wastes, handled in 55 gallon steel drums for con-
venience, are deposited in our secure landfill. The drums of waste
are entombed in the bulk sludge generated by the primary treatment
portion of the WWTP. Approximately 15,000 drums per year are
handled.

     The secure landfill is constructed in three adjoining five-
acre segments, the second of which is currently active while the
third is just being constructed. Construction provides a cleared,
prepared, sand covered base, graded to a one percent slope toward
one side. The entire segment is surrounded by a dike of specially
selected impermeable compacted clay. The bottom and sides of the
basin so created are first lined with reinforced "Hypalon"

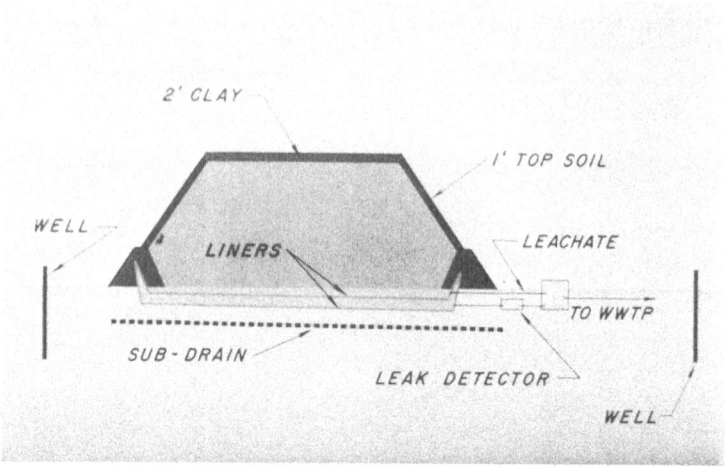

Fig. 3. Schematic Drawing of Secure Landfill.

(chlorosulfonated polyethylene) (Fig. 3). A second liner of
"Hypalon" is installed, separated from the first by pea-sized
gravel overlain by a layer of protective sand. The inter-spacial
area serves as a leak detector and assures against any significant
hydraulic head on the bottom liner. On top of this uppermost liner
a one foot layer of pea-sized gravel is applied, covered by a sheet
of "Typar" (spun bonded polypropylene). The "Typar" serves to keep
the subsequently applied sludge from clogging the gravel, thus
ensuring free flow of leachate to the sump from which it is pumped
to the WWTP.

The sludge from the WWTP, which is 40-50 percent solids, has
physical characteristics similar to "dry" sandy soil. It is this
sludge in which the drummed wastes, previously referred to, are
buried.

When the maximum elevation of the adjoining segments is reached,
a final closing, with two feet of compacted impervious clay, is
applied, covered with top soil and seeded to protect against erosion.

Throughout construction, operation and beyond, samples from
monitoring wells which completely surround the landfill area are
scrutinized for any indication of intrusion of landfill contaminants into the ground water.

The third component of our waste management facilities is the
WWTP (Fig. 4). Since this is the most unique of our facilities,
I will discuss it in some greater detail.

Fig. 4. Aerial View of Waste Water Treatment Plant,
Chambers Works, E. I. du Pont de Nemours.

Waste Water Treatment Plant

The Chambers Works Waste Water Treatment Plant disposes of a
waste stream having a range of characteristics: (Table 1)

The organic component of the waste varies from simple acids
and alcohols, to substituted aromatic compounds, to polymeric compounds. The composition and quantity of waste varies considerably
with time. For example, the organic carbon could range from 100 to
300 milligrams per liter from one day to the next. The make-up one
day could be high in low molecular weight alcohol and the next day
high in nitro aromatic compounds.

Technology Development

In planning and designing the Waste Water Treatment Plant, the
nature of the waste stream posed a major challenge to the engineers
working to meet the deadline of an NPDES Permit to be enforced in
mid 1977. Activated sludge alone did not do the job on color or
organic carbon and suffered frequent toxic upsets. Physical-chemical
means were proposed but were expensive, too specific, and not particularly effective. An activated sludge plant in conjunction with
the carbon columns did not work well and would have required
excessive investment.

Table 1. Waste Water Treatment Plant

Influent

Flow	20,000 - 25,000 gpm
Acidity	250 - 1,680 mg/liter
Total BOD$_5$	60 - 240 mg/liter
Organic Carbon (OC)	122 - 209 mg/liter
Color	430 - 2,000 APHA
Total Suspended Solids (TSS)	150 - 460 mg/liter
Total Dissolved Solids	1,400 - 6,000 mg/liter

Note: The low ratio of BOD/OC indicates a waste relatively
difficult to treat by biological oxidation.

At that point the team undertook a novel experiment and combined powder-activated carbon with activated sludge. The immediate
result was that the combination significantly outperformed the
activated sludge and PACT* was invented. Subsequent laboratory work
showed that the PACT Process would meet the limits of the NPDES
Permit and offer several corollary benefits. (Table 2)

Plant Design and Operation

In primary treatment (Fig. 5), the waste water is first
compacted with lime slurry in three, parallel 200,000 gallon
neutralizers. Overflow from the neutralizers passes through four
clarifiers in parallel. A polyelectrolyte is added to enhance
settling prior to the clarifiers. Each rectangular clarifier is
one million gallons in volume. Traveling bridges rake the solids
to the inlet end of the clarifier where an underflow slurry of

Table 2. Powdered Activated Carbon Treatment (PACT)

Advantages

Operate Without Toxic Shock Upsets
Consistently Remove Over 90 Percent of the BOD
Remove Over 85 Percent of the OC
Reduce Color
Eliminate Foaming
Significantly Improve Sludge Settling and Filtration

---

*Registered U. S. Service Mark for Du Pont's Waste Water Treatment
Service.

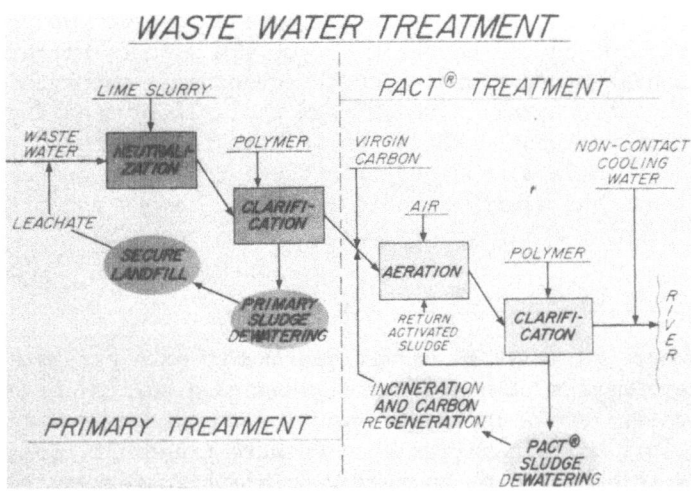

Fig. 5.   Schematic Flow Sheet of Waste Water Treatment Plant -
          Chambers Works, E. I. du Pont de Nemours.

five to ten percent solids is removed to a filter feed tank. High
pressure filters are then used to dewater the sludge to a 45 to 50
percent solids level. Each filter generates about twelve thousand
pounds per cycle. The cycles may be as short as an hour to an hour
and one-half and reach a final filtration pressure of just over 200
pounds per square inch. This yields approximately 120,000 pounds
per day of filtered primary solids generated for disposal in a
secure landfill.

     The effluent from primary treatment at a pH of 6.0 to 8.0 is
fed to three parallel four million gallon aerators. A mix of
regenerated and virgin activated carbon is added to the primary
effluent at an average dosage of 120 milligrams per liter. An
underwater system of air piping disperses air throughout the
aerators. The air flow of thirty to fifty thousand standard cubic
feet per minute per aerator is sufficient to suspend the mixed
liquor suspended solids (MLSS) and supply the necessary oxygen for
bacterial oxidation. The treated waste and the MLSS flow into two
parallel 2.5 million gallon clarifiers. Underflow recycled activated sludge from the clarifiers is returned to the aerators at a
recycle ratio of 0.5. The treated water overflows the clarifiers
and is mixed in the final basin with non-contaminated cooling water.
It is then discharged to the Delaware River estuary.

Wasted biological/carbon mass is processed through a sludge
thickener and a filter. The design of the filter is the same as
the one described for primary. Filter cake at forty to forty-five
percent solids is fed into the top of a twenty-five foot diameter
by a forty foot high multiple-hearth furnace. There are five
hearths in the furnace. The rising hot gases evaporate water from
the sludge entering on the top hearth. On the middle, hotter hearths
the biomass and absorbed organics on the carbon are pyrolyzed. On
the lower, hottest hearths in the presence of water vapor, the spent,
powdered carbon is regenerated for recycle to the aerators. The
incandescent carbon is discharged from the furnace and quenched in
water. The resultant slurry is acid washed to remove ash prior to
recycle to prevent a build-up of inorganics in the MLSS.

Plant Performance of the PACT Process

The Chambers Works NPDES Permit has twenty-nine parameters
including heavy metals, ammonia, total suspended solids, organic
carbon, pH, color, etc. The performance of the plant is determined
by the control of certain variables. Primary treatment, of course,
is controlled to meet the pH and heavy metal parameters in the
permit. In the PACT part of the operation, carbon dosage and air
flow are adjusted to treat the varying organic components of the
feed and meet the effluent permit limits. Continuous dissolved
oxygen meters and carbon analyzers provide the required control data.

Table 3. Average Waste Water Treatment Plant Performance
          1978 thru 1980

Flow	24,800 gpm
Effluent Quality	Average Value
Soluble BOD5, mg/liter	6.7
Percent Removal	96
DOC, mg/liter	31.5
Percent Removal	82
Color, APHA	475
Percent Removal	67
Carbon Dose, mg/liter	120
Percent Virgin	54
Percent Regenerated	46

Table 3 shows the performance of the PACT operation in the years 1978 through 1980. During this three-year period, the plant met its discharge permit limits for BOD, organic carbon and color, the key parameters controlled by PACT, more than 99.5 percent of the time.

The startup of the PACT process and its operation demonstrated two unique concepts in waste water treatment:

- It achieved performance equal to tertiary treatment in a two-stage process.

With the exception of the regeneration furnace system, the type of equipment which was installed for the PACT process is identical to that which would be used in an activated sludge process. An important aspect of the PACT process is that it is readily adaptable to a conventional activated sludge plant. The carbon improves the sludge settling characteristics and clarifiers operate at higher flux ratio.

The PACT process is not nearly as sensitive to toxic upsets as activated sludge because the activated carbon protects the biomass. Actually, there is a synergistic effect between the two which enhances the biological activity.

- The regeneration of powdered-activated carbon from PACT sludge was demonstrated in a multiple hearth furnace. Although multiple hearth furnaces have been used for years, this is the only instance where we know that this type of equipment is used for powdered carbon regeneration. The quality of the carbon regenerated in the furnace as measured by iodine absorption is generally fifty percent of that of initial virgin carbon.

One of the important features of the PACT process is that the excess activated sludge is incinerated in the furnace during the carbon regeneration.

In summary, Table 4 shows the operating conditions for the Chambers Works PACT process.

Table 4. Operating Conditions

PACT Process

Sludge Age	20 - 40 Days
MLSS	15,000 $\pm$ 20,000 mg/l
Aerator Dissolved Oxygen Level	1 - 4 ppm
pH	6.5 - 8
Temperature	25° - 30°C (77° - 100°F)
Carbon Dosage	100 - 150 ppm

Recent Laboratory Studies

During the development and in the initial stages of operation,
the performance of the PACT process was assessed by removal of conventional classes of pollutants.

More recently the activated sludge process and the PACT process
were compared for organic priority pollutant removal. In this test,
a single batch of feed waste water was used instead of the constantly
variable feed to the plant. Both units were operated with eight-hour
retention time and a ten-day sludge age. NUCHAR SA 15 carbon was
used in two concentrations, 100 ppm and 300 ppm. Results for volatile
organics (Table 5) showed little difference in performance, both
giving good removal.

Table 5.   Comparison of PACT and Activated Sludge for
           Volatile Organics Removal

(Laboratory Results)

Compound	Feed Concentration ppb	Activated Sludge	% Removal By PACT 100 ppm*	PACT 300 ppm*
Benzene	81	98.5	99.6	99.6
Chlorobenzene	3,660	99.1	99.7	99.8
Chloroethane	667	99.8	>99.9	>99.9
Chloroform	72	96.7	97.2	96.9
Methylchloride	138	98.5	>99.7	>99.7
Tetrachloroethylene	33	>99.5	>99.5	>99.5

*Carbon dose expressed as ppm

However, the PACT process significantly improved removal of
base-neutral extractable organic compounds (Table 6), giving increased removal with increased carbon dosage. For acid extractable
compounds, the PACT process also performed significantly better
than activated sludge.

Table 6. Comparison of PACT and Activated Sludge for
Extractable Organics Removal

(Laboratory Results)

Compound	Feed Concentration ppb	Activated Sludge	% Removal By PACT 100 ppm*	% Removal By PACT 300 ppm*
Base-Neutral Extractables				
1,2 - Dichlorobenzene	18	90.6	97.8	>99
2,4 - Dinitrotoluene	1,000	31	60	90
2,6 - Dinitrotoluene	1,100	14	94	95
Nitrobenzene	330	94.5	99.5	>99.9
1,2,4 Trichlorobenzene	210	>99.9	>99.9	>99.9
Acid Extractables				
2,4 - Dichlorophenol	19	0	84	93
2,3 - Dinitrophenol	140	39	97	>99
4 - Nitrophenol	1,100	25	91	97

ACKNOWLEDGMENTS

The author thanks Mr. George Cassedy and Mr. Harry Heath for
valuable assistance in the preparation of this presentation.

A REGIONAL OFF-SITE HAZARDOUS WASTE MANAGEMENT FACILITY

IN PRACTICE

M.W. Hunt

Rollins Environmental Services (NJ) Inc.
Bridgeport, New Jersey, United States

SUMMARY

Successful operation of a regional hazardous waste treatment facility has proven to be feasible. Some important facts about an incineration process operating as part of such a facility are:

1) Each waste must undergo pre-acceptance qualifications.

2) Contractual agreements minimize liability questions.

3) Effective quality control measures ensure proper treatment.

4) Pre-incineration treatment is an integral step in proper waste disposal.

5) Effective incinerator instrumentation aids in the attainment of adequate waste destruction efficiency.

6) Incinerator residuals must be disposed of in an environmentally sound manner.

7) Keys to safe, effective regional treatment facilities are sound management coupled with practical environmental regulations.

8) Regional waste treatment facilities provide very significant benefits in the form of treatment capability and economy, resulting in important environmental benefits.

INTRODUCTION

"Regional" describes a concept widely practiced for the purpose
of providing such things as services for education, sewage treatment, water supply, transportation, etc. It had not been practiced
to any great extent as an approach to hazardous waste management,
however, until the past two decades. Even then, the majority of
the early attempts usually involved only very low level technology.

The reason for this was primarily that there were few, if any,
environmental regulatory requirements in effect until after 1972.
Since off-site treatment would normally denote treatment by someone
other than the waste generator and, therefore, would result in an
additional operating expense, there was little incentive for construction of such facilities.

In the mid-1960's, RLC Corporation, which included Matlack
Trucking, a major bulk petro chemical transporter, concluded that
such type of waste treatment facility could be justified. These
facilities could treat the wastes generated from clean-out of the
bulk transports and also accept wastes from other generators who
lacked on-site treatment capability.

These plans came to fruition with the construction and startup of RLC's first regional facility at Bridgeport, New Jersey
in 1970. While a number of such facilities were introduced in Europe
in the mid-1970's, it is our current understanding that this was
the world's first regional facility directed exclusively to concentrated, liquid industrial wastes, which we now call Hazardous
Wastes, equipped with full-scale incineration capability. Two
major, similar facilities followed in 1971 at Baton Rouge,
Louisana, and Deer Park, Texas (Houston area). All three of
these facilities were provided with capabilities for incineration,
chemical neutralization, biological treatment, some metal recovery
and limited or extensive secure landfill. The Baton Rouge
facility also grew to include landfarming and later added a deep
well satellite facility about 30 miles away.

These three facilities were to be part of fifty (50) such
facilities throughout the United States. Unfortunately, due to
the early pre-regulatory timing and resultant poor return on
investment, further expansion became impossible to pursue at that
time. The original facilities have continued to operate since then,
although many changes have been made in order to respond to the
various regulatory requirements as well as to the waste generators' needs.

This paper describes the current operations of the incinerator system as practiced at the Bridgeport, New Jersey facility.
The only significant differences between this incinerator and the

other two are the state environmental standards in effect, the operational costs where subject to geographic variation, and types of waste available for treatment within the regional market. The purpose of this paper, however, is to describe the most significant aspects regarding the operation of a regional hazardous waste incinerator in general, rather than to address these differences. The Rollins facility is a privately owned and operated commercial facility which provides contracted services to hundreds of generators. As a result of this, various contractual and business elements of the operation will be addressed as well as the technical aspects of operations.

GENERAL OPERATIONS

Pre-Acceptance Qualification of the Waste

Once the waste generator has determined that they have material that is of no commerical value to them as a processor, such material is classified as a waste. The term hazardous is added if it meets certain criteria specified by the Federal Government's Resource Conservation and Recovery Act of 1976. Generally, the definition either identifies the waste by chemical name, generating process, or by one of four categories—ignitability, corrosivity, reactivity or toxicity. The Rollins facility is accessible for any waste that is compatible with our system and whose generator wishes to contract it; all wastes, however, are subjected to the Hazardous Waste Control System, whether they are so defined or not.

The first action required of the generator, therefore, is to establish the key characteristics of their waste. If they contract Rollins, such information is documented by use of a waste data sheet (Figures 1A and 1B) and submitted to our Technical Department so it can be determined if and how the waste can be treated. There are several key pieces of data to be included on this form, some of which are mandated by law, such as the EPA Identification No., concentration of mercury, lead, sulfur, chlorinated compounds, and BTU value. Other information is needed to ensure compliance with state regulations or ensure operation within process limitations.

Once the treatability decision is made and concurred with by the operating department, a treatment price is calculated by the Technical Department. This price is unique for that particular waste and includes limitations on the variability of key parameters such as heat content, ash content, and halogenated content. The price must then be approved by operations, sales and the Vice President.

The sales representative then presents a proposal to the prospective customer based upon the above information. Normally

this proposal specifies a check burn volume that must be evaluated
prior to a final proposal.

No wastes are priced nor are any wastes accepted at any Rollins
plant until information is submitted to Rollins as specified on
the Waste Data Sheet. If the generator cannot provide such
information or chooses not to, Rollins will agree to provide the
necessary analytical service at pre-established prices.

Container Type	Price Range
Bulk liquid (40,000 lbs)	$.05 to $.40 per pound
Drums (55 gallon steel drums)	$60 to $400 each
Drums (20 gallon to 40 gallon fiber drums)	$40 to $300 each
Pharmaceuticals (misc. package size)	$.30 to $.40 per pound.

Prices for disposal by incineration vary significantly as
a function of the wastes' key characteristics. This variation is
far more pronounced than for secure landfill, as an example,
because variable costs are directly related to heat incineration
content, halogen content, ash content and throughput rate. In
addition, fixed costs have steadily increased as a result of
regulatory-mandated capital requirements such as secondary containment, more stringent emission standards, continuous monitoring and
instrumentation. Generally, current incineration prices at the
Bridgeport, New Jersey facility range as follows:

WASTE DATA SHEET | ROLLINS
                                    | ENVIRONMENTAL SERVICES

CUSTOMER INFORMATION:
COMPANY NAME:_____RES STREAM NO._____
PLANT ADDRESS:_____MAILING ADDRESS:_____
STATE_____ STATE:_____ ZIP:_____
COMPANY CONTACT, TECHNICAL:_____ PHONE:_____
COMPANY CONTACT, BUSINESS:_____ PHONE:_____
USEPA GENERATOR I.D. NO._____STATE GENERATOR I.D. NO.____

GENERAL WASTE DESCRIPTION:

Type of Process Generating Waste:_____

Quantity Generated (per mo.) |Frequency (of removal)

TRANSPORTATION INFORMATION:

Hazardous Material:

Hazardous Substances	Concen-tration	Hazardous Substances	Concen-tration

Hazardous Characteristics:_____

Transporter:_____   Placarding:_____

TRANSPORTATION EQUIPMENT:

Tank Truck []   Vacuum Truck [] Flatbed [] Dump Truck []
Bin   []   Barge   [] Tank Car [] Other []   ____

Method of Collection

Fiberpaks []   Drums [] Tanks [] Sumps [] Other[] _____

Other available transportation information:_____

FIGURE 1A

DETAILED WASTE DESCRIPTION AND REGULATORY COMPLIANCE:

RCRA Characterization Codes ____ ____ ____ ____
Reason for above characterization:_____
State Characterization Codes ____ ____ ____ ____
OSHA: Contain Listed compounds? ____   EPA: PCB Conc <50 ppm?___
NRC: Radioactive? ____   PHS: Infectious Wastes?_____
FIFRA: Does this waste contain a pesticide for which the
        EPA has issued specific disposal requirements? _____

CHEMICAL COMPOSITION:

Compound Name	Norm. Conc. Range % W	Chemical Formula

LABORATORY ANALYSIS | PHYSICAL PROPERTIES

Metals

PHYSICAL STATE @25°C

Metals					PHYSICAL STATE @25°C	
Pb	___Mg/L	CN	___Mg/L	GAS __	LIQUID __	
Hg	___Mg/L	TOC	___Mg/L	SOLID __	SLUDGE __	
Cd	___Mg/L	COD	___Mg/L	SLURRY __	PASTE __	
Be	___Mg/L	BOD	___Mg/L	GANULAR __	CRYSTAL __	
As	___Mg/L	SS	___Mg/L	POLYMERIC __	AMOSPHOUS __	
Na/K	___Mg/L	TDS	___Mg/L			
Other	___Mg/L	Br	___Mg/L	SINGLE PHASE ___	BTU___/lb.	
	___Mg/L	Cl	___Mg/L	MULTI PHASE ___	ASH___%	
	___Mg/L	F	___Mg/L	OIL/WATER ___	VAPOR PRESS	
	___Mg/L	I	___Mg/L	VISCOSITY ___	@___	
	___Mg/L	S	___Mg/L		MELTING PT __	
					BOILING PT __	
					pH __	
					FLASH PT __	

Is the waste reactive with water? _____ with air? _____
Is a representative sample provided?_____
Give any other additional information on the hazards of the
waste:_____

I hereby certify that the above information is complete and
accurate.

Customer Signature   Title   Date

FIGURE 1B

Contractual Agreements

The treatment proposal is accompanied by a standard contract form (Figures 2A, 2B and 2C). If the customer accepts the treatment price, Rollins prefers that this acceptance be documented by returning a signed contract along with a purchase order. In some cases, more involved contracts are negotiated in order to meet special customer needs:

Key portions of the standard contract are:

* Indemnification - Specifies when transfer of waste ownership occurs (Rollins, by preference, transports most of the wastes between the generator and the treatment site and, therefore, as owner of the waste upon pick-up, accepts the transportation liability).

* Warranties - Addresses needs for proper regulatory compliance by Rollins and proper waste identification by generator.

* Force Majeure - Protects either party against uncontrollable situations.

* Payment Terms - Requires timely payment.

* Safety and Health Conditions - Requires that both parties must have specific knowledge regarding known carcinogens as required by the Federal Occupational Safety and Health Act.

In summary, use of such a contract minimizes potential liability difficulties by clearly defining the limits of the business relationship.

The specifics of each waste stream are handled by an Appendix so that detail modifications can be readily handled without disturbing the nature of the basic relationship.

DATE:_____
RES Ref. NO._____

This letter, upon receipt by Rollins Environmental Services (NJ) Inc. ("RES") of your acceptance shall be the agreement between RES and   ("Company") with respect to Waste (defined below), term price and representations:

WARRANTY-RES. To comply with all existing laws, ordinances and regulations of the United States and of any state, county, township or municipal subdivision thereof, or other governmental agency which may be applicable to the removal of Waste, if RES provides transportation, as well as the processing and/or treatment of the Waste. RES shall obtain all permits, licenses, and other forms of documentation required in order to comply with such laws and regulations.

RES INDEMNIFICATION. Following loading and departure from Company's plant, if RES provides transportation or, following delivery F.O.B. RES' facility, if Company provides transportation, Company shall be relieved of responsibility and RES shall become solely responsible for any and all loss, damage or injury to persons or property and RES shall indemnify and hold Company harmless from any and all liability, damages, costs, claims, demands and expenses of whatever type of nature, including, but not limited to, pollution or other damage, which shall be caused by, arise out of, or in any manner be connected with the Waste, except as provided in COMPANY INDEMNIFICATION below.

COMPANY WARRANTS. Company represents and warrants that the Waste loaded and removed under this Agreement shall be the Waste defined on Schedule "A", attached hereto and made a part hereof, and has been thoroughly characterized on the waste data sheet submitted to RES. Company agrees to prepare and execute RES' waste data sheet for each shipment of Waste. If the Waste is packaged, Company warrants that such Waste shall be prepared for shipment and packaged in containers specified by the then current and applicable regulations of the Unites States Department of Transportation, Environmental Protection Agency or any successors thereto and/or any state, municipal and/or Federal agency having jurisdiction, as the case may be. Company shall be responsible for loading packaged Waste on RES' trailers if RES is providing transportation.

COMPANY INDEMNIFICATION. Company will indemnify and hold harmless RES from any and all loss, damages, including damage or undue wear and tear to equipment, claims, suits or costs which shall arise or grow out of any injury to any person or persons or any property (including the person or property of Company or its employees) caused by or resulting in any way from Company's

FIGURE 2A

failure to comply with Company's Warranty concerning any and all
liability, damages, costs, claims, demands, and expenses of
whatever type or nature resulting from the acts and/or omissions
of Company and/or its employees, until departure of RES vehicles
from Company's plant, if RES provides transportation or, if
Company provides transportation, until delivery F.O.B. RES'
facility.

    1. <u>TERM</u>. Subject to the right of either party to terminate
this Agreement at any time upon thirty (30) days prior written
notice, this Agreement shall automatically terminate on

_____.

    2. <u>PAYMENT</u>. RES shall invoice Company for the hauling and
treatment of Waste at the rates and terms set forth on Schedule
"A" attached hereto and made part hereof. RES shall add an
amount equal to one and one-half percent ($1\frac{1}{2}$%) or the maximum
legally permissible amount to invoices which remain unpaid for
more than thirty (30) days after date of invoice. Like charges
may be made for each subsequent thirty (30) day period that such
invoice remains unpaid.

    3. <u>FORCE MAJEURE</u>. Delays or failures of either party in
the performance of its required obligations shall be excused if
caused by circumstances beyond the reasonable control of the
party affected, including, but not limited to, acts of God,
strikes, labor holiday, fire, flood, windstorm, explosion, riot,
war, sabotage, action or request of governmental authority,
accident, inability to obtain material, equipment or transportation, provided that a prompt notice of such delay is given
and the parties shall be diligent in attempting to remove such
cause(s).

    4. <u>OSHA</u>. Company represents and warrants that Waste does
not contain the following substances in concentrations greater
than those specified below:

2-acetylaminofluorene, Chemical Abstracts Service	
Registry No. 62759	1%
alpha-naphthylamine, Chemical Abstracts Service	
Registry No. 132327	1%
4-aminodiphenyl, Chemical Abstracts Service	
Registry No. 92671	0.1%
benzidine, Chemical Abstracts Registry No. 92875	0.1%
beta-naphthylamine, Chemical Abstracts Service	
Registry No. 91598	0.1%
beta-propiolactone, Chemical Abstracts Service	
Registry No. 57578	1%
bis-chloromethyl ether, Chemical Abstracts	
Service Registry No. 542881	0.1%

FIGURE 2B

3,3'-dichlorobenzidine, Chemical Abstracts Registry No. 91941 and its salts	1%
4-dimethylaminoazobenzene, Chemical Abstracts Service Registry No. 60117	1%
ethyleneimine, Chemical Abstracts Service Registry No. 151564	1%
methyl chloromethyl ether, Chemical Abstracts Service Registry No. 107302	0.1%
4,4'-methylene bis (2-chloroaniline), Chemical Abstracts Service Registry No. 101144	1%
4-nitrobiphenyl, Chemical Abstracts Service Registry No. 92933	0.1%
N-nitrosodimethylamine, Chemical Abstracts Service Registry No. 62759	1%
polychlorinated biphenyls	0.005%

Additions may be made by RES to the foregoing list of substances from time to time, such additions by RES becoming effective
and binding after three days' written notice to Company.

Company agrees that all Waste containing asbestos (including actinolite, amosite, anthophyllite, chrysotile, crocidolite,
and tremolite) fibers longer than 5 micrometers detectable by
phase contrast microscopy shall be subject to the following conditions:

(a) The presence of asbestos in Waste shall be clearly
noted on RES' waste data sheet.
(b) Waste shall be packaged in closed steel drums bearing a label which conforms with 29 CFR 1910.1001.

Company further represents and warrants that, to the best
of its knowledge, Waste does not contain vinyl chloride monomer in a
liquid or gaseous form except as specified on RES' waste data
sheet.

All previous representations, including but not limited to
proposal(s) purchase order(s) and/or invoice(s), either written
or oral are hereby annulled and superseded. No modification shall
be binding unless in writing and executed by RES and Company.

Please indicate your agreement to the above recitals by
executing and returning a copy of this letter.

ACCEPTED THIS ____ day of ____, 19___ .   ROLLINS ENVIRONMENTAL
                                            SERVICES (NJ) INC. ("RES")

_____("Company")   By:_____
By:_____   Address: P.O. Box 221
Address: Bridgeport, NJ 08014

FIGURE 2C

Waste Delivery Quality Control

If the customer agrees to this proposed business relationship, each of the individual wastes are assigned a unique, four
digit number, consecutive in nature from the first day of operation,
with a prefix which initiates the creation of a unique set of
documentation which thereafter controls all future handling of
that specific waste stream.

A Waste Safety Sheet (Figures 3A and 3B) is jointly prepared
by the Safety and Technical Departments. Prior to any acceptance
of such waste, a copy of that specific Waste Safety Sheet must be
provided to the following:

* Truck Driver - must have a copy of such for whatever
  waste he is transporting.

* Security - must have a safety sheet for every waste on
  site and identified by waste location. This is important
  in case of an emergency. A copy, for the waste involved,
  would accompany an injured person in case of chemical
  exposure which required off-site medical treatment.
  This same information would be made immediately available
  to community emergency response personnel, such as a
  fire department.

* Laboratory - must have a complete set of safety sheets
  for all wastes handled.

* Safety Supervisor - must have a complete set for emergency reference.

* Unloading/Storage Control Room - must have a complete set
  of safety sheets. However, only those for the wastes
  which are presently in storage or being unloaded must be
  visibly displayed and identified as to waste location.

* Incinerator Control Room - must have a complete set of
  safety sheets. However, only those presently being
  fed to the incinerator must visibly be displayed and
  identified as to waste location.

WASTE SAFETY SHEET | ROLLINS
                                            | ENVIRONMENTAL SERVICES
Waste Designation:_____
Rollins #_____ EPA#_____ STATE#_____

CHEMICAL COMPOSITION:

COMPONENT NAME	FORMULA	RANGE W %	FLASH PT. °F

PHYSICAL PROPERTIES:

Physical Character:_____
Color:_____ Odor:_____
Flash point [] <100°F [] <140°F
                    [] >140°F

INSPECTION FOR RECEIPT APPROVAL:

Parameter	Min. - Limits - Max.

Treatment Method/Process Notes:
_____
_____
_____
_____

☐ There is a special procedure
   for this waste

HAZARD CODE INFORMATION

4=Red-Severe 3=Orange-High

2=Yellow-Moderate 1=Blue-Low

CHEMICAL COMPATIBILITY:
_____
_____
_____
_____
_____

REQUIRED PERSONAL PROTECTIVE EQUIPMENT:

[x] Hard Hat   [ ] Safety Glasses   [ ] Air-Supplying
[x] Rubber Gloves   [ ] Splash Goggles   Respirator
[ ] Rubber Boots   or Face Shield   [ ] Air-Purifying
[ ] Protective Suit   [ ] Other_____   Respirator

                                                Type:_____

_____   _____
Date Completed   Authorized Signature

FIGURE 3A

Waste Designation:_____ Rollins #_____

HEALTH PROTECTION:
  Effects of Exposure:_____

  First Aid: Eyes:_____
              Skin:_____
              Inhalation:_____
              Ingestion:_____
  Notes for Physician:_____

FIRE PROTECTION:
  Explosion or Fire Hazard:_____
  Fire Fighting Procedures:_____

REACTIVITY
  Conditions to Avoid:_____
  Incompatibles:_____
  Hazardous Products:_____

SPILL RESPONSE PROCEDURES:_____
_____
_____

TRANSPORTATION INFORMATION:

HAZARDOUS MATERIAL:

Hazardous Substances	Concentration	Hazardous Substances	Concen-tration

Hazardous Characteristics:_____
_____
_____

Placarding:_____

FIGURE 3B

When the waste leaves the generator, a hazardous waste
manifest must be initiated by the generator and must accompany the
transporter.   Figures 4A and 4B comprise an example of such a manifest as required by New Jersey and subsequently revised to meet
RCRA requirements.   It now consists of a two part, six copy form.
This type of manifest is intended to provide the "cradle-to-grave"
documentation for every hazardous waste shipment throughout the
United States.   It permits tracking of each hazardous waste from
the generator through every step of handling and requires reporting
of any deviation from the norm up to and including final disposal.

Distribution of the various copies are:

- Three (3) copies of Part A are left at generator's
  site after waste pick-up.

Upon delivery of the waste to Rollins and laboratory acceptance of same, remaining copies are distributed as follows:

- One (1) copy of Part B to the New Jersey Department
  of Environmental Protection (site of the disposal
  facility).

- One (1) copy of Part B to the generator's state.

- One (1) copy of Part B to the generator.

- If the waste was transported by someone other than
  Rollins, that company would receive a transporter's
  copy of the manifest.

- All other copies of both parts are retained by Rollins
  for at least three (3) years.

Upon arrival at the treatment facility, each waste must be
sampled and analyzed for key parameters to verify that it is definitely the same waste as was originally characterized by the waste
data sheet, waste safety sheet and manifest.   If the laboratory
confirms same, the manifest is signed and the waste unloading
personnel are instructed, in writing, as to the required disposition of that waste.   This includes unloading instructions
(specific receiving tank) and type of treatment to be done.

If the waste delivery does not conform to specifications,
the laboratory immediately determines if the out-of specification
waste can be properly treated at the Rollins plant.   If it can be
treated, but at a higher cost, or if it cannot be treated due to
process limitations, the sales representative advises the generator
of the new situation.   The generator then specifies as to the disposition of same.   If the price revision is acceptable, the material

STATE OF NEW JERSEY
DEPARTMENT OF ENVIRONMENTAL PROTECTION
HAZARDOUS WASTE MANIFEST

PART A: SEND TO DISPOSER'S STATE DOCUMENT NO. NJ 0106256

Generator Name	Phone (inc. area code)	EPA I.D. No.
Address (Street-City-State-Zip Code		
Transporter No. 1	Phone (inc. area code)	EPA I.D. No.
Transporter No. 2	Phone (inc. area code)	EPA I.D. No.
Address (Street-City-State-Zip Code		
Treatment, Storage or Disposal (TSD Facility Phone EPA I.D. No.		

IF MORE THAN TWO TRANSPORTERS ARE TO BE UTILIZED, FILL OUT THE
FOLLOWING AS APPROPRIATE
THIS FORM IS NO__OF A TOTAL OF__. THE FIRST MANIFEST DOCUMENT
NO. IS NJ_____

PROPER US DOT SHIPPING NAME	US DOT HAZ-ARD CLASS	UN NO.	Form	NET Qty.	Units	Containers No.	Type	EPA Haz. Code	EPA Waste Type
1.									
2.									
3.									
4.									
5.									

Special Handling Instructions Including Container Exemption (i.e.
Identification of Additional Wastes Included in Shipment of a Non-
Hazardous Nature which do not have to be manifested)

GENERATOR'S CERTIFICATION: This is to certify that the above
named materials are properly classified, described, packaged,
marked and labelled and are in proper condition for transportation according to the applicable regulations of the Dept. of
Transportation, USEPA and the State. The wastes described above
were consigned to the Transporter named. The Treatment, Storage
or Disposal Facility can and will accept the shipment of hazardous waste and has a valid permit to do so. I certify that the
foregoing is true and correct to the best of my knowledge.

Generator's Signature Title	Date Shipped	Exp. Arr. Date
Treatment storage signature & certi Title fication of receipt of shipment	Date Received	

FIGURE 4A

PART B: SEND TO DISPOSER'S STATE   GENERATOR'S EPA I.D. NO.

Transporter No. 1 Signature and Certification of Delivery and Non-Tampering with Shipment Also Print Signature	Date Delivered
Transporter No. 2 signature and certification of Receipt of Shipment – Also Print Signature	Date Received
Transporter No. 2 SWA Registration No.	
Transporter No. 2 Signature and Certification of Delivery and Non-Tampering with Shipment Also Print Signature	Date Delivered
Treatment Storage or Disposal Facility of any differences between Manifest and Shipment or Listing of Reasons for and Disposition of Rejected Materials TSD FACILITY EPA ID NO.	Handling Method 1.____ 4.____ 2.____ 5.____ 3.____ 6.____
Treatment Storage or Disposal Facility Title Signature & Certification of Receipt of Shipment–Also Print Signature	Date Received

In case of emergency or spill immediately call the
State the Emergency occurred in and the NJ Dept.
of Environmental Protection
(609) 292-5560 (day)   (609) 292-7172 (night)

DOCUMENT NO. NJ 0106256

FIGURE 4B

is unloaded for treatment. If the price revision is rejected
or if treatment is not possible, the generator specifies the
return delivery location and absorbs all resultant costs under
previously specified contractual conditions.

In cases of such rejections, same must be noted on the
manifest. In all cases, Rollins is responsible for distribution
of the manifest copies as described above.

This entire quality control procedure is extremely critical
in order to ensure proper compliance with safety and environmental requirements as well as to fulfill all mutual contractual
obligations.

Pre-Incineration Treatment Procedures

Additional standard operating procedures (SOP's) must be rigidly followed once the waste has been accepted by Rollins for disposal. Part of these are mechanical type safety/environmental procedures such as:

* Determination as to receiving vessel or location.

* If bulk liquid delivery, a fail safe electrical ground and nitrogen vapor space must be established on the delivery trailer.

* Proper pre-specified safety gear must be worn.

* Determination of compatibility of new waste with wastes already in storage.

The other key procedures include pretreatment steps as appropriate such as filtration, pH adjustment phase separation and blending. All burnable liquid wastes (>2,000 BTU/lb.) are blended so as to prepare a feed tank of waste (typical volume of about 20,000 gallons which will sustain the liquid waste burner system for about 20 to 24 hours) with the necessary heat content (7,500 to 10,000 BTU/lb.) to maintain operating temperatures. Other key parameters to be controlled by proper blending include:

* Compatibility

* Halogen content

* Sulfur content

* Ash (heavy metal) content

Incineration Controls

Figure 5 shows a schematic of the Rollins incinerator as currently operated. This paper will not describe the design details because of their complexity. However, copies of a paper given earlier in 1980 by a Rollins staff member on that subject are available for distribution here today.

Since this incinerator is designed to burn a combination of up to four (4) types of waste at any one time (kiln solids, high BTU liquids, aqueous liquids and gases), the first priority regarding incinerator control is to ensure that all wastes are burned at an adequate temperature to accomplish chemical destruction. That temperature varies somewhat (assuming that residence time and excess oxygen levels are fairly stable).

Rollins presently maintains a relative high minimum temperature of
2000°F for all non-PCB wastes. Continuous positive control is
accomplished by an automatic waste feed shutoff system which is
actuated by a thermocouple which measures the temperature at the
hot duct.

There are other instrumentation interlock systems which
are equipment protection devices such as automatic fan shutdowns
in case of excessive vibration, loss of cooling in the scrubbing
system, etc. These systems are also critical in order to achieve
safe incinerator operation.

Assuming that the earlier procedures have been followed,
the next critical step is to make a check burn with any new waste
so as to evaluate its treatability from a control and throughput
standpoint. This is <u>not</u> the test burn procedure mandated by
RCRA, however, since those regulations have not been promulgated
yet, but instead consists of carefully monitored incineration of
limited quantities of a single waste (i.e. six fiber pak drums
to the kiln). Upon completion of the check burn, the other controls
reviewed earlier are normally adequate to achieve proper incineration.

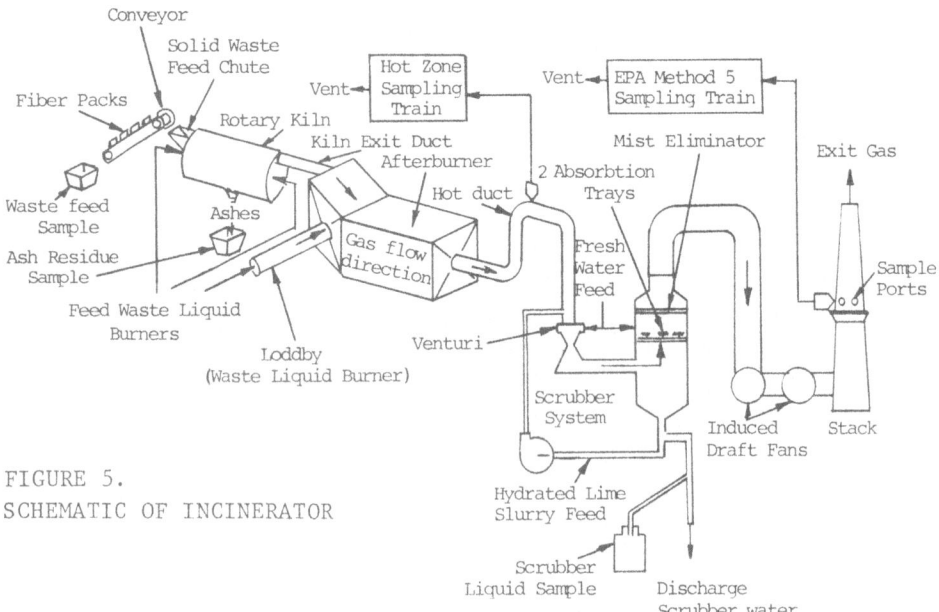

FIGURE 5.
SCHEMATIC OF INCINERATOR

Residual Disposal

Waste incineration produces three types of residuals that must be disposed of in an environmentally sound manner:

- Kiln ash – This material has been water quenched and consists of inorganic ash from the discharge end of the kiln as well as the metal rings and lids from the fiber drums. Although EP leachate tests indicate that this material is not hazardous by definition (test leachate must contain<100 times parameters specified in drinking water standards), a petition for delisting has not been submitted yet, so all such materials must be disposed of in a secure chemical landfill. Current generation rate is about seven (7) cubic yards per day.

- Combustion gas – Based upon the incinerator design and documented performance, chemical destruction efficiencies exceed 99.99%. Acid gas and particulate content must be controlled in the emission scrubber system, however, by use of gas cooling, alkaline scrubbing and impingent separators. Typical scrubber performance is shown in Table I. The purified combustion gas is then discharged through a stack to the atmosphere as a water-saturated gas at about 140° to 165°F.

TABLE I

INCINERATOR EMISSION CONTROL

Parameter	Lbs./hr.	
	Emissions w/o controls	Emissions with controls
SO_x	550	120
Hcl	4500	50
NO_x	150	75
Particulate	200	0.1 gr/DSCF*

*Grains/dry standard cubic foot corrected to 12% CO_2

- Emission Scrubber Sludge – the combustion gas scrubber system generates a lime sludge which flows continuously to a series of settling lagoons where clarification occurs. The clarified water is of adequate purity (see Table II) to permit direct discharge (under permit) to an adjacent tidal stream. This is done on

a batch basis coincident with high tide to enhance
transport to the Delaware River from the creek into
which the discharge flows.

The sludge is periodically dewatered and removed for
disposal to a secure chemical landfill. This material
also appears to pass the EP leachate tests regarding
toxicity, but a delisting petition has not yet been
submitted. Current generation rate is about ten (10)
cubic yards per day at 40% solids content.

TABLE II

Incinerator Effluent Discharge*

Parameter	Average Concentration (mg/l)
BOD_5	<15
COD	<50
Oil and grease	<9
Suspended Solids (total)	<30
Dissolved Solids (total)	<5400
Ammonia (as nitrogen)	<20
Aluminum	<1
Arsenic	<0.05
Barium (soluble)	<1
Beryllium	<0.05
Cadmium (total)	<0.02
Chromium (hexavalent)	<0.05
Copper	<0.1
Lead	<0.15
Mercury	<0.005
Molydenum (total)	<1
Titanium	<2
Zinc (total)	<0.3
pH	6-9
flow (maximum)	1,120,000 gal/day

*This data is based upon actual discharge data, however, has been
been corrected to adjust for contribution from other sources
such as the biological treatment process.

Administrative Controls

The previously described systems and/or procedures are all
extremely important in themselves, but a safe, environmentally
sound, efficient waste treatment facility can only be attained
through sound management practices, coupled with practical
environmental regulations.

Due to the rather limited opportunities for people to gain experience in hazardous waste management because of the small number of such operations in existence for only the past few years, staffing is somewhat challenging. The Bridgeport facility census consists of about eighty-five (85) people (including sales personnel), total, approximately half of whom are supervisory and clerical. About 25% to 30% of the total are degreed professionals, specializing in engineering, chemistry or industrial safety and health.

Management must then ensure that all personnel are adequately trained so that each person will instinctively and effectively perform according to the following priorities:

A) Facility personnel and community safety.

B) Environmental compliance.

C) Process security.

D) Profitability improvement.

In order to accomplish this objective, it is necessary to rely heavily upon written standard operating and emergency response procedures as well as effective inspection and preventive maintenance programs, supported by a cadre of experienced personnel.

CONCLUSIONS

Use of regional hazardous waste treatment facilities are not only technically feasible as evidenced by more than thirty (30) total operating years within the three regional facilities of Rollins Environmental Services, Inc., but definitely offer significant benefits to the industrial community and public at large.

This approach promotes optimal productive use of capital funds for the industrial community. This service also enhances the potential for effective regulatory enforcement because compliance inspection efforts, etc. can concentrate on fewer operating locations.

By actual practice, it has been demonstrated that regional facilities offer waste generators access to technologies and economies of large-scale processing which they would unlikely be able to provide for themselves.

Both of these advantages indirectly benefit the public by keeping the economy more competitive world-wide through effective use of capital and enhancement of environmental improvement.

THE BAVARIAN HAZARDOUS WASTE SYSTEM, ILLUSTRATED BY THE

EBENHAUSEN TREATMENT PLANT AND GALLENBACH LANDFILL SITE

Franz Defregger

Bavarian Ministry for Environmental Affairs

Munich, Federal Republic of Germany

SYSTEM DESCRIPTION; INSTITUTIONS

A long time before the Waste Disposal Acts were drafted
in the Federal Republic of Germany, Bavaria started to
build up a country-wide system for the disposal of special waste. Because of the size (area: 70.500 sq.km, inhabitants: 11 million) of Bavaria and her economical
structure, this system contains several central disposal
facilities and collecting (transfer) stations. At present,
10 regional collecting stations and 3 central treatment
plants are in operation in Bavaria (Schweinfurt, Schwabach,
Ebenhausen/Gallenbach). There had been treated almost
350.000 tons of special waste in 1980 (see Table 1).

With the exception of the treatment plant of Schwabach,
which is run by a municipal co-operative, called "Zweck-
verband Sondermüllplätze Mittelfranken," all the other
facilities and collecting stations are run by the "Ge-
sellschaft zur Beseitigung von Sondermüll in Bayern mbH"
(Company for Disposal of Special Waste in Bavaria ltd.),
for short "GSB", which was established as a country-wide
organization in 1970. Its stock fund amounts arose from
1 Million in 1970 to 21 Millions DM today. Shareholders
of the company are the Bavarian State (78 %), 3 municipal organizations (8 %) and 56 industrial Companies (14
%). The task of the GSB is to provide and operate facilities necessary for the treatment of special waste and
the recovery of raw materials from special waste all
over Bavaria. The activity of the company is conducted

Table 1

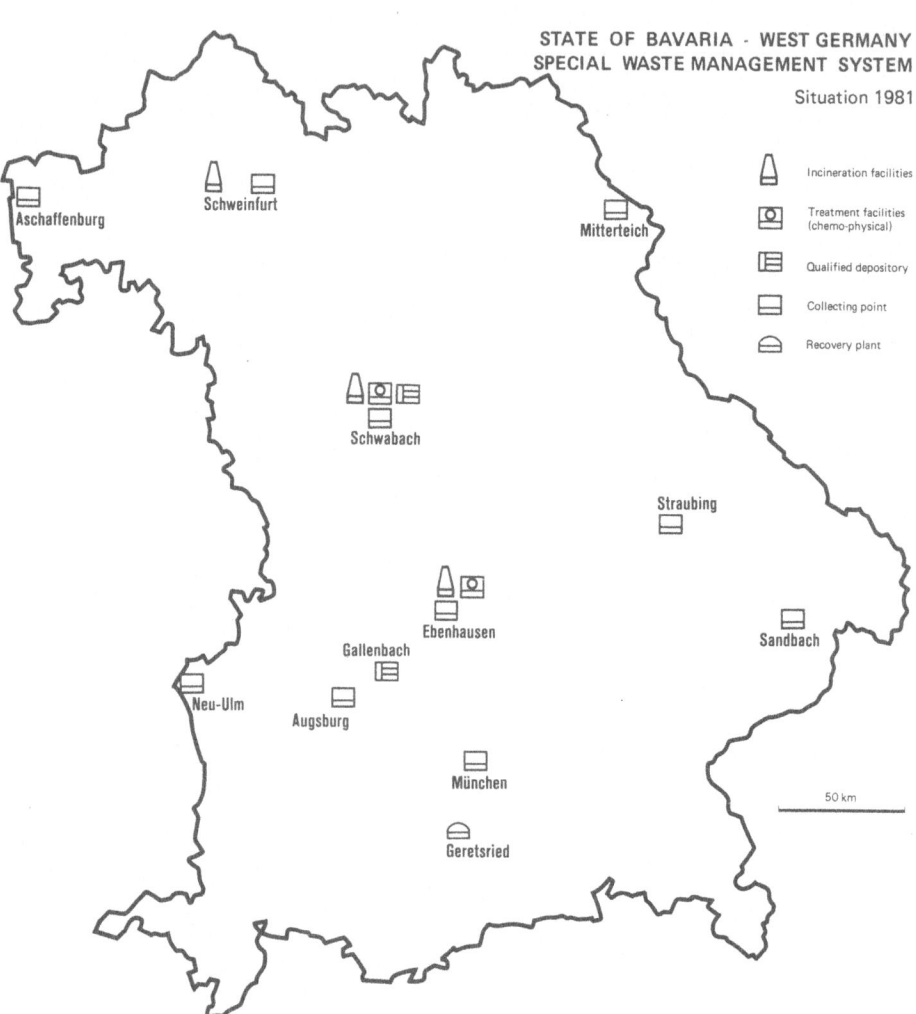

STATE OF BAVARIA · WEST GERMANY
SPECIAL WASTE MANAGEMENT SYSTEM

Situation 1981

on a not-for-profit basis, but being a private company the
GSB has to search permanently new ways to treat the hazardous waste in order to get a proper disposal or a recovery of raw materials from this waste.

COLLECTION AND TRANSPORTATION

The special waste is temporarily deposited at the collecting points, where it is classified for bulk transport to the central disposal plants. The gathering
points of GSB have simple pre-treatment facilities for
separating oil-water mixtures into water and solids, and
for neutralizing and sludge-thickening.

Waste collection and transportation is controlled by a
complex numbering system of invoices to be filled in and
signed by the waste generator who keeps one copy as a
record. The shipper keeps one, too, while handing over
the other copies, the order form and confirmation when
he delivers the waste at the gathering point (or treatment plant) where the materials are weighed and a quality is sampled to verify the shippers record.
This trip-ticket-System is based on federal law and
controlled by the Bavarian Environmental Protection
Agency and the Local Authorities.

EBENHAUSEN CENTRAL TREATMENT PLANT

The biggest and most modern of the three Bavarian hazardous waste facilities is the facility Ebenhausen/Gal-
lenbach which started up 1976 on a 4-hectare site in
Ebenhausen near Ingolstadt and a 17-hectare landfill in
Gallenbach 25 miles away (no suitable site was available
next to the treatment plant Ebenhausen). The Ebenhausen
plant comprises a laboratory, a chemical-physical treatment plant for organic and inorganic substances (oil-
emulsions, used acids, alkalies, galvanizing and other
inorganic sludges, solutions containing chrome, cyanides, nitrites, etc.), a water purification plant, as well
as Germany's largest waste incineration plant with a capacity of 100 times 10 to the sixth power btu/h (see
Table 2).

The incineration plant, able to process 70.000 tons/yr of
waste, has two parallel rotary kilns for solid and semi-
solid wastes and a common burner-chamber with a set of six
burners for liquid wastes. The heat of the off-gases from
the after-burners is utilized in a steam boiler where the
off-gases are cooled down from 1.000 centrigrades to about

Table 2

PLAN OF SITE

Scale 1:1250

Ebenhausen Central Treatment Plant

(incineration and chemo-physical treatment)

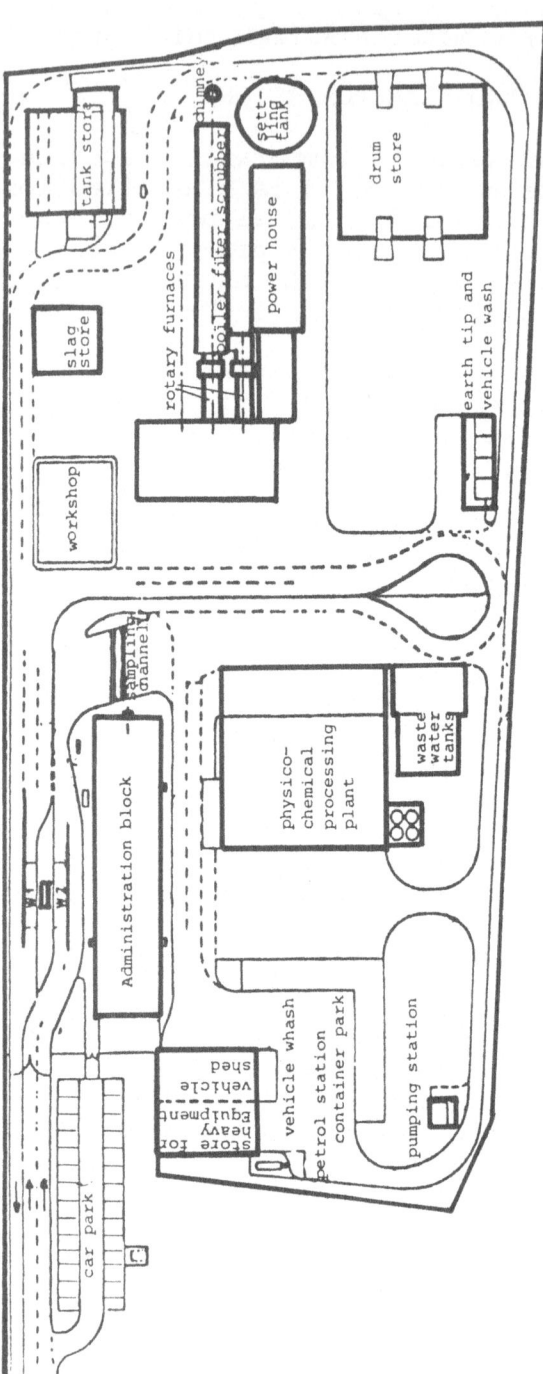

270 centigrades, thereby generating up to 30 tons steam/h
out of which 22 tons/h are consumed in a 1,530 kw steam
turbine while the remainder is condensed in an air conden-
ser; this electric energy not only supplies the incineration plant's entire power requirement, but actually leaves
an excess supply for the public grid. The steam from the
turbine (three atomospheres pressure) is utilized for heating the building and for process heat in the chemical-physical treatment plant.

Off-gas purification comprises an electrical precipitator for dust retention and a two-stage venturi-type
scrubber which removes HCl and HF almost completely while
retention of SO_2 is 70 pc.

There are also scrubbers for treating the gases and waste
air from the storage tanks of the chemical-physical treatment plant, so as to avoid odors. The sludge from this
plant, as well as from a GSB-developed facility for treating specifically oil- and emulsion-carrying industrial
waste waters (two basins, a centrifuge, a mixing and
dosage measuring device and a vacuum drum filter), is de-
watered in sludge presses and the resulting solid materials is land-filled in Gallenbach.

GALLENBACH LANDFILL SITE

General Description

The Gallenbach special waste site covers an area of 17
hectares which slopes down about 30 m to the west and at
a depth of 10-20 m is underlain by a layer of clay loam
8-18 m thick. On the basis of these favourable hydrogeological conditions, seeping of water into the lower ground
water level is prevented. Above the loam there are
moraine deposits such as gravel and sand, which were removed to construct the site.

In contrast with conventional landfill sites, here the
site had to be prepared by excavation and construction
of dams. The soil thus excavated, and otherwise in short
supply, is employed to cover the deposited special waste.

The site has a capacity of about 1.4 million m³; this
corresponds to a tipping period of about 20 - 25 years.
Construction is planned in 4 phases, the first of which
is for a deposition volume of 450.00 m³, started end of
1975 and is still in operation (see Table 3)

Table 3

PLAN OF SITE: Scale: 1:2500

Gallenbach Landfill Site

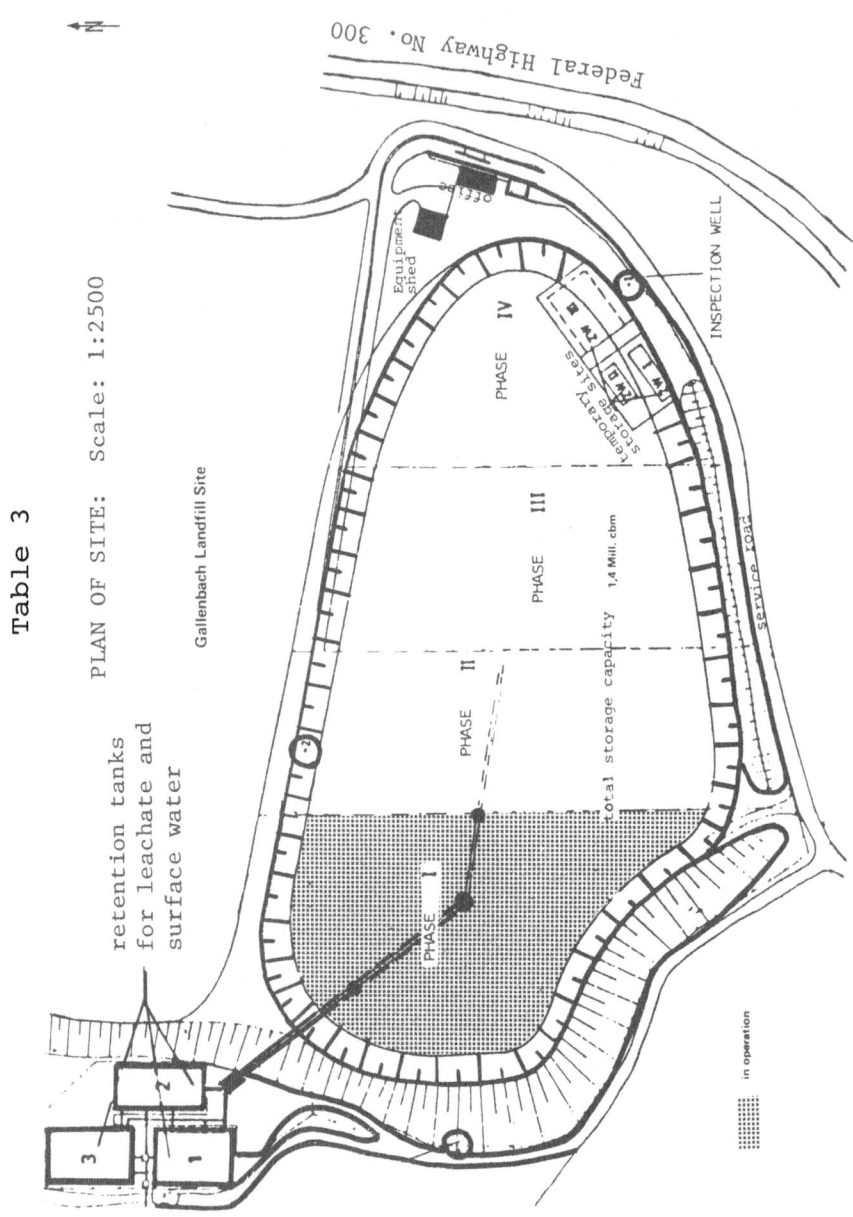

On the basis of the official authorisation the following special types of waste can be deposited at the site:

- Combustion residues from the incinerators of the GSB

- Soil contaminated with oil.

- Hydroxide sludges which have been detoxified, neutralised and de-watered satisfactorily.

- Process-specific wastes already pretreated at the gathering points of the GSB or delivered direct to the site.

- Other solid production waste not disposed of together with domestic waste and capable of deposition in the site without preliminary treatment.

A record must be kept of the waste deliveries, giving the source, volume, type and composition of the waste in each case.

The plant comprises the following installations (see Table 3):

- Operational building with amenity rooms of various types for the personnel.

- Laboratory for study of the substances delivered.

- Vehicle weigh bridge.

- Vehicle and instrument shop.

- Control systems (drainage, retaining basins) for holding and treating the surface and seepage water (leachate).

There is also one earth basin sealed with foil, erected as intermediate store for sludges containing oil. This sludge is treated and/or burned subsequently in the Ebenhausen plant.

Collecting, analysing and treating of leachate and surface water (research programs)

Although the thick loam clay under the site virtually eliminates penetration of surface and seepage water into the deeper groundwater zones, a loam seal at least 40 cm thick was also applied to the bottom of the site and on

the inner side of the dams. The seepage water is collected by drainage systems (perforated plastic pipes,
diameter 10 cm) in filter gravel packing, and led via
seepage water shafts (diameter 10cm) into a main mani-
fold (diameter 25cm, concrete tubes) to seepage-water
basins below the site.

Independently of this the surface water is dealt with in
its own system. Both types of water reach a total of 3
retaining basins (tanks) with a capacity of about
10.000 m³ via sand tramps. These tanks are sealed with a
2 mm thick plastic foil (Schlegel panel). The water from
each filled tank is monitored currently. The surface
water has no contamination and is led into the receiving stream, the Paar, a tributary of the Danube. The
leachate from the retaining basins is pumped into lorry
and transported to the chemical physical treatment
plant in Ebenhausen.
The quantity of lechate generated in the landfill, section I (catchemt area : 30.000 m²) during 1976 - 1980
shows the following Table 4:

Table 4: Quantity of leachate 1976 - 1980 in section I
(catchment area : 30.000 m²)

Year	Precipi-tation (mm)	Surface water (m³)	Leachate (m³)	%	Delivered solid wastes (m.to)
1976	530	16.000	14.000	87	60.250
1977	830	25.000	15.100	60	66.270
1978	774	24.300	13.300	54	62.170
1979	900	26.400	12.000	45	60.480
1980	698	21.000	8.200	39	71.740

Since 1976, the quantity of leachate could be reduced
from over 80 % to nearly 40 % in 1980 by subjecting each
layer of waste to a high degree of compaction and by
applying layers of impermeable material (landfilling
only in small sections 30 m x 30 m).

Ever since the landfill has been on operation, the leachate and groundwater have been controlled and analysed by
the Bavarian Environmental Protection Agency. The average
analytical results of leachate of the last years are shown
in Table 5:

Table 5: Average leachate concentrations (mg/l) in
         Gallenbach 1977 - 1980

Parameters	1977	1978	1979	1980
pH	7,0	7,4	7,2	7,4
BOD_5	2720	3000	4000	4650
COD	3000	5500	9000	12000
TOC	1600	1900	4000	3400
Cr	0,4	0,7	0,5	0,2
Cu	0,3	0,3	0,4	0,3
Cn	-	-	0,1	0,1
Ni	4,1	5,2	3,6	4,4
Zn	2,3	0,5	0,6	0,4
Cd	0,2	0,1	0,2	0,2
Fe	12	4	4	2,1
Sn	2,3	8	44	34
Pb	0,6	1,0	1,2	1,1
Phenole	2,1	7,3	1,0	10
Chloride	15000	26000	37000	40000
Sulfate	2100	3400	8300	3300

The leachate is characterized by high organic (BOD, COD,
TOC) and inorganic concentrations, in particular salt
contents between 15 g/l and 40 g/l and ammonia between
30 mg/l and 500 mg/l, also some heavy metals like zinc
with 40 mg/l. In view of the high contamination of
the leachate, it is pumped from the retaining basin into
lorry and after transport treated in the chemo-physical
plant in Ebenhausen. A new research program is underway now, treating the leachate in a one- and two-stage-
distillation pilot plant. First results give the impression that TOC and COD could be reduced over 90 %.
The distillate is free of inorganic salts and heavy metals. The contaminated residues of this distillation are
dry and crystalline. The research program will be finished by the end of 1982.

The following Table 6 shows the annual quantity of
organic and inorganic contamination as well the amount
of heavy metals.

Table 6: Annual Quantity (in kg) of organic and inorganic material in leachate of Gallenbach

year	1976	1977	1978	1979	1980	Summary 1976 - 1980
Leachate (m^3)	14000	15100	13300	12000	8200	62600
TOC	7000	24200	26200	48200	27900	133900
BOD	22100	41000	41800	48500	38100	191500
Mineraloil	1001	147	41	44	328	10561
Phenole	168	32	88	12	88	388
Chloride	124000	234000	361000	444000	320000	1483000
Sulfate	7000	31000	48000	99000	27000	212000
Ammonia	732	1770	6062	10800	7380	26744
Chromium	-	6	10	7	2	25
Iron	140	178	43	47	17	425
Nickel	46	62	73	43	36	260
Copper	3	5	4	5	3	20
Zinc	28	35	7	7	3	80
Cadmium	5	3	2	2	2	14
Zinn	-	116	644	537	284	1581
Lead	20	9	14	14	9	66
						2471 kg

From nearly 62600 m³ leachate, generated 1976-1980, over
1480 metric tons were produced as salt and about 2470 kg
as heavy metals.

Last but not least, for continuous monitoring of the
groundwater, observation wells are installed around the
site, from which samples are taken at monthly intervals
for analysis as regard various parameters (BOD$_5$, COD,
O$_2$-content, chloride, sulphate, heavy metals). Since
opening the landfill site in 1976, there is no contamination of groundwater.

(Editor's Note: This paper was solicited by the Symposium
but the author was unable to present it
to the Symposium. It is included as an
addendum to the proceedings, because of
its pertinence and interest to the Symposium participants. --JPL).

CONTRIBUTORS

Mdme. Jacqueline Aloisi de Larderel
Director, Solid Waste Program
Ministere de l'Environnement et
   Cadre de Vie
14 Boulevard du General Leclerc
92521 Neuilly-sur-Seine, France

Mr. Jack Bentley
Superintending Chemist
Land Waste Division
U.K. Department of the Environment
1, Lambeth Palace Road
London SEI 7ER, England

Mr. Ken Childs
Chief, Technology Transfer Division
Environment Canada
Ottawa, Ontario KIA I C8, Canada

Mr. Eugene Crumpler (WH-565)
Assessment & Technology Branch
Hazardous & Industrial Waste Division
U.S. Environmental Protection Agency
401 M Street, S.W.
Washington, D.C. 20460

Mr. Franz Defregger
Bavarian Ministry for Environmental Affairs
Munich, Federal Republic of Germany

Dr. C.J. Duyverman
Organization for Applied Scientific Research (TNO)
The Netherlands

Dr. A. Goudsmit
Director, Solid Waste and Clean
  Technologies Division
Ministry of Health and Environmental
  Protection
Postbus 439
2260 AK Leidschendam, The Netherlands

Mr. Maurice Hunt
Vice President
Rollins Environmental Services
1 Rollins Plaza
Wilmington, Delaware 19899

Dr. Gunnar Johnson
Director, Kali and Salz A.G.
Friedrich Eberstrasse 160
P.O. Box 102029
D 3500 Kassel, Federal Republic of Germany

Mr. John P. Lehman (WH-565)
Director, Hazardous & Industrial Waste Division
Office of Solid Waste
U.S. Environmental Protection Agency
401 M Street, S.W.
Washington, D.C. 20460

Mr. Pierre Lieben
Environment Directorate
Organization for Economic Cooperation
  and Development
2 rue Andre Pascal
75775 Paris Cedex 16, France

Mr. Etienne Le Roy
Agence Nationale Pour la Recuperation
  et l'Elimination des Dechets
2, Square la Fayette
B.P. 406
49004 Angers, Cedex, France

Dr. Ted E. Lewis
General Superintendent
E.I. Du Pont, Chambers Works
Box 174
Rockland, Delaware 19732

Mr. Jean M. Massin
Director, Water Pollution Prevention Service
Ministere de l'Environnement et du
  Cadre de Vie
14 Boulevard du General Leclerc
92521 Neuilly-sur-Seine, France

Mr. Christian Nels
Assistant Head of Section
Thermal Treatment
Federal Environmental Agency
Bismarckplatz 1
D1 Berlin 33, Federal Republic of Germany

Dr. Kurt Riegel (RD-681)
Associate Director
Office of Environmental Engineering & Technology
Office of Research and Development
U.S. Environmental Protection Agency
401 M Street, S.W.
Washington, D.C. 20460

Mr. Benno K. Risch
European Community
Rue de la Loi 200
B 1049 Brussels, Belgium

Mr. Alan I. Roberts
Associate Director for Hazardous Materials
  Regulation
U.S. Department of Transportation
400 Seventh Street, S.W.
Washington, D.C. 20590

Mr. Norbert Schomaker
Chief, Disposal Branch
U.S. EPA Municipal Environmental
  Research Laboratory
26 W. St. Clair Street
Cincinnati, Ohio 45268

Dr. Robert Stephens
California Department of Health Services
2151 Berkeley Way
Room 235
Berkeley, California 94704

Mr. Axel Szelinski
Director, Regulatory Affairs
Federal Environmental Agency
Bismarckplatz 1
D1 Berlin 33, Federal Republic of Germany

Mr. J. Toffner-Clausen
Managing Director
Kommunekemi
Lindholmvej 3
DK-5800 Nyborg, Denmark

Mr. Knut Trovaag
Director, Norcem A/S
3470 Slemmestad, Norway

Dr. Fritz Van Veen
Infra Consult
The Netherlands

Mr. Per Waage
Deputy Director General
Pollution Control Department
Ministry of Environment
Myngaten 2
P.O. Box 8013, DEP.
N-Oslo 1, Norway

Dr. Bernd Wolbeck
Ministry of Interior
Graurheindorfer Strasse 198
53 Bonn 1, Federal Republic of Germany